BTEC

DIFFERENTIAL CALCULUS

AND APPLICATIONS

ANTHONY NICOLAIDES
B.Sc (Eng.) C. Eng. M.I.E.E.
Senior Lecturer and Course Tutor
for the BTEC NATIONAL DIPLOMA IN ENGINEERING

at

LEWISHAM COLLEGE

P.A.S.S. PUBLICATIONS

PRIVATE ACADEMIC & SCIENTIFIC STUDIES LIMITED

©   A. Nicolaides 1991

First Published in Great Britain 1991 by

Private Academic & Scientific Studies Limited.

ISBN 1 872684 08 4

Other titles pulished by the same author in the BTEC series.

Algebra III
Mathematics II
Analytical Mathematics II
Electrical and Electronic Principles II
Calculus III

Other titles published by the same author in the GCE A series.

Complex Numbers
Trigonmetry
Algebra
Cartesian and Polar Curve Sketching

Printed by Hartnolls Limited
A member of Martins Printing Group
Victoria Square, Bodmin, Cornwall, PL31 IEG

PREFACE

This book, which is part of the G.C.E. A level series in Pure Mathematics covers the specialised topic of Differential Calculus and Applications.

The A level series in Pure Mathematics is comprised of 10 books covering the syllabuses of most boards. The books are designed to assist the student wishing to master the subject of Pure Mathematics.

Each book deals extensively with a specialised topic with ample examples worked out in full. The series is easy to follow with minimum help.

The Differential Calculus and Applications, like all the books in the series, is divided into two parts. In part I, the theory is comprehensively dealt with, together with many worked examples and exercises. A step by step approach is adopted in all the worked examples. Part II of the book, a special feature acts as a problem solver for all the exercises set at the end of each chapter in Part I.

I am grateful to Sandra Francis for typesetting the manuscript on a desktop publishing system.

Thanks are due to the following examining bodies who have kindly allowed me to use questions from their passed examination papers.

University of Cambridge Local Examinations Syndicate.   C

The Associated Examining Board.   AEB

The University of London School Examinations.   U.L.

The University of London School Examinations Board accepts no responsibility whosoever for the accuracy or method of working in the answers given.

DIFFERENTIAL CALCULUS AND APPLICATIONS

CONTENTS

PART I

1. **ALGEBRAIC FUNCTIONS.**

   Differentiation from first principles.   1, 3

   The idea of a limit and the derivative defined as a limit. The gradient of a tangent as the limit of the gradient of a chord.   2

   THE DERIVATIVE OF A SUM OR DIFFERENCE OF A FUNCTION.   9

   THE DERIVATIVE OF A PRODUCT.   11

   THE FUNCTION OF A FUNCTION.   16

   IMPLICIT FUNCTIONS.   18

2. **TRIGONOMETRIC OR CIRCULAR FUNCTIONS.**   25

   The derivative of sin x from first principles.   25

   The derivative of cos x from first principles.   25

   The derivative of tan x.   26

   Derivative of:-

   a) Sum or difference rule.   30
   b) Product rule.   26
   c) Quotient rule.   35
   d) Function of a function.   35
   e) Implicit functions.   35

   Inverse trigonometric functions.   35

3. **EXPONENTIAL FUNCTIONS.**   40

$y = a^x, y = 2^x, y = e^x$

The derivative of e^x from first principles.   42

4. **LOGARITHMIC FUNCTIONS.**   48

5. **HYPERBOLIC FUNCTION.**   54

Inverse hyperbolic functions.   59

6. **PARAMETRIC EQUATIONS.**   65

7. **SECOND AND HIGHER DERIVATIVES OF A FUNCTION.**   76

8. **TANGENTS AND NORMALS.**   83

9. **SMALL INCREMENTS AND APPROXIMATION RATES.**   91

10. **NEWTON-RAPHSON'S METHOD.**   104

Methods of approximation to the solution of an equation, improvements of
the value of such approximation including the
Newton-Raphson method.   114

11. **MACLAURIN'S EXPANSION.**

Third and higher derivatives. Quadratic and higher degree polynomial
approximations to simple functions.

PART II

SOLUTIONS 1   136
SOLUTIONS 2   153
SOLUTIONS 3   160
SOLUTIONS 4   167
SOLUTIONS 5   172
SOLUTIONS 6   182
SOLUTIONS 7   188
SOLUTIONS 8   194
SOLUTIONS 9   198
SOLUTIONS 10   206
SOLUTIONS 11   211

MISCELLANEOUS UNIVERSITY EXAMINATION QUESTIONS.   222

DIFFERENTIAL CALCULUS AND APPLICATIONS

DIFFERENTIATION

1.   **ALGEBRAIC FUNCTIONS.**

Differentiation from first principles. The idea of a limit and the derivative defined as a limit. The gradient of a tangent as the limit of the gradient of a chord.

Consider the general form

$$y = a\,x^n$$

where a and n are constants and x and y are variables, x is the independent variable and y is the dependent variable.

If $a = 1$ and $n = 2$ then

$$y = x^2$$

the graph takes the form of a parabola as shown in **Fig. 1** for $x \geq 0$.

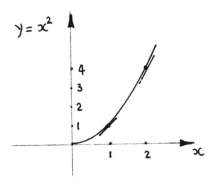

Fig. 1

It is required to find the graident at a point, say, $x = 2$. A tangent is drawn at $x = 2$ as shown, the gradient is defined as the ratio of the vertical

distance, 4, to the horizontal distance, 1, that is, gradient $= \dfrac{4}{1} = 4.$

The gradient at $x = 1$, is again found as $\frac{2}{1} = 2.$

The method of finding gradient at different point of a curve is rather approximate and tedious. A much neater method is used in finding gradients, <u>differential calculus</u> method.

Consider a point **P** on the parabola.

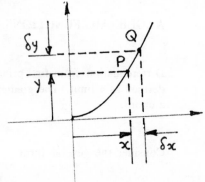

Fig. 2

as shown in **Fig. 2** of coordinates **P** (x, y), a point **Q** close to **P** has coordinates **Q** $(x + \delta x, y + \delta y)$ where δx and δy are infinitecimally small distances along the x and y axes respectively.

Expanding **Fig. 2** for convenience and clear illustration as in **Fig. 3**. The chord **PQ** is drawn, the gradient

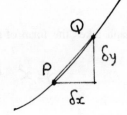

Fig. 3

of the chord **PQ** is given as $\delta y/\delta x$, that is, gradient of **PQ** $= \delta y/\delta x$.

CONCEPT OF A LIMIT

Consider a length of 1 metre, halving this length gives a length of 500 millimetres, halving this again, gives a length of 250 mm, this, gives a length of 125 mm, keep on halving the lengths indefinitely the length will approach a very small length such as $\delta x \to 0$ (delta x tends to zero).

If we apply this concept to **Fig. 3** when **Q** approaches **P** and $\delta x \to 0$ then the gradient of the chord **PQ** will approach the gradient of the tangent at **P**

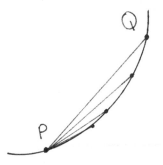

Fig. 4

gradient at **P** $= \left(\dfrac{\delta y}{\delta x}\right)_{\delta x \to 0}$

gradient at **P** $= \dfrac{dy}{dx}$

$\dfrac{dy}{dx}$ is the ratio $\dfrac{\delta y}{\delta x}$ as δx tends to zero.

NOTATION OF THE GRADIENT

The gradient of a function is denoted as $\dfrac{dy}{dx}$.

This has alternative terminology, the <u>gradient</u> of the tangent at **P**, <u>the slope</u> of the tangent at **P**, the <u>first derivative</u>, or the <u>primitive</u> of the function.

TO DERIVE THE FIRST DERIVATIVE OF THE FUNCTION $y = x^2$ FROM FIRST PRINCIPLES

If x is increased to $x + \delta x$, then the corresponding value of y is $y + \delta y$.

$$y + \delta y = (x + \delta x)^2 \dots(1)$$
$$y = x^2 \dots(2)$$

subtracting (2) from (1)

$$\delta y = (x + \delta x)^2 - x^2$$
$$\delta y = x^2 + 2x\, \delta x + \delta x^2 - x^2$$
$$\delta y = 2x\, \delta x + \delta x^2$$

dividing each term by δx

$$\frac{\delta y}{\delta x} = 2x + \delta x \ldots (3)$$

applying the idea of the limit, as $\delta x \to 0$ $\quad \frac{\delta y}{\delta x} \to \frac{dy}{dx}$ in (3), we have

$$\boxed{\frac{dy}{dx} = 2x}$$

TO DERIVE THE FIRST DERIVATIVE OF THE FUNCTION $y = 2x^3$ FROM FIRST PRINCIPLES

If x is increased to $x + \delta x$, then the corresponding value of y is $y + \delta y$

$$y + \delta y = 2(x + \delta x)^3 \ldots (1)$$

$$y = 2x^3 \qquad \ldots (2)$$

substracting (2) from (1)

$$\delta y = 2(x + \delta x)^3 - 2x^3$$
$$\delta y = 2(x^3 + 3x^2\, \delta x + 3x\, \delta x^2 + \delta x^3) - 2x^3$$
$$\delta y = 2x^3 + 6x^2\, \delta x + 6x\, \delta x^2 + 2\, \delta x^3 - 2x^3$$
$$\delta y = 6x^2\, \delta x + 6x\, \delta x^2 + 2\, \delta x^3$$

dividing by δx each term, we have

$$\frac{\delta y}{\delta x} = 6x^2 + 6x\, \delta x + 2\, \delta x^2 \ldots (3)$$

applying the idea of the limit, as $\delta x \to 0$, $\quad \frac{\delta y}{\delta x} \to \frac{dy}{dx}$ in (3), we have

$$\boxed{\frac{dy}{dx} = 6x^2}$$

From the last derivations, it can be deduced that the derivative of $y = a\,x^n$ is

$$\boxed{\dfrac{dy}{dx} = an\,x^{n-1}}$$

Observe that the coefficient of ax^n is multiplied by the power n and the power is reduced by unity.

$$y = x^2, \quad \frac{dy}{dx} = 1 \times 2\,x^{2-1} = 2x$$

$$y = 2\,x^3, \quad \frac{dy}{dx} = 2 \times 3\,x^{3-1} = 6\,x^2$$

$$y = ax^n \quad \frac{dy}{dx} = anx^{n-1}.$$

WORKED EXAMPLE 1

Derive from first principles the derivatives of the functions:-

(i) $y = 3x$ (ii) $y = 2\,x^2$ (iii) $y = 5\,x^3$

(iv) $y = 5/x$ (v) $y = 3\,x^{1/2}.$

SOLUTION 1

(i) $y = 3x$ ...(1)
$y + \delta y = 3\,(x + \delta x)$ ...(2)
(2) - (1)

$\delta y = 3\,(x + \delta x) - 3x$
$\delta y = 3x + 3\,\delta x - 3x$

$\delta y = 3\,\delta x$

dividing each side by δx

(ii) $y = 2\,x^2$ ...(1)
$y + \delta y = 2\,(x + \delta x)^2$ ...(2)
(2) - (1)
$\delta y = 2\,x^2 + 4x\,\delta x + 2\,\delta x^2 - 2\,x^2$

$\delta y = 4x\,\delta x + 2\,\delta x^2$
dividing each term by δx, we have

$\dfrac{\delta y}{\delta x} = 4x + 2\,\delta x$

as $\delta x \to 0, \quad \dfrac{\delta y}{\delta x} \to \dfrac{dy}{dx}$

5

$$\delta y / \delta x = 3$$

$$\boxed{\frac{dy}{dx} = 4x}$$

as $\delta x \to 0$, $\quad \delta y / \delta x \to \dfrac{dy}{dx}$

$$\boxed{\frac{dy}{dx} = 3}$$

(iii) $\quad y = 5\,x^3 \qquad\qquad\qquad \ldots(1)$

$y + \delta y = 5\,(x + \delta x)^3 \ \ldots(2)$

$\delta y = 5\,(x + \delta x)^3 - 5\,x^3 = 5\,(x^3 + 3\,x^2\,\delta x + 3\,x\,\delta x^2 + \delta x^3) - 5\,x^3$

$\delta y = \delta x^3 + 15\,x^2\,\delta x + 15\,x\,\delta x^2 + 5\,\delta x^3 - 5\,x^3$

$\delta y = 15\,x^2\,\delta x + 15 \times \delta x^2 + 5\,\delta x^3$

dividing each term by δx

$$\frac{\delta y}{\delta x} = 15\,x^2 + 15 \times \delta x + 5\,\delta x^2 \text{ as } \delta x \to 0, \ \frac{\delta y}{\delta x} \to \frac{dy}{dx} \quad \frac{dy}{dx} = 15\,x^2$$

(iv) $\quad y = 5/x \ \ldots(1) \qquad y + \delta y = \dfrac{5}{x + \delta x} \ \ldots(2) \qquad \delta y = \dfrac{5}{x + \delta x} - \dfrac{5}{x}$

$$\delta y = \frac{5x - 5\,(x + \delta x)}{x\,(x + \delta x} = \frac{5x - 5x - 5\,\delta x}{x\,(x + \delta x)} \qquad \delta y = \frac{-5\,\delta x}{x\,(x + \delta x)}$$

dividing each term by $\delta x \quad \delta y / \delta x = \dfrac{-5}{x\,(x + \delta x)}$ as $\delta x \to 0$, $\delta y / \delta x \to dy/dx$

$$\frac{dy}{dx} = -\frac{5}{x^2}.$$

Alternatively, $\quad y = \dfrac{5}{x} = 5\,x^{-1} \ \ldots(1) \ y + \delta y = 5\,(x + \delta x)^{-1} \ \ldots(2)$

(2) - (1)

$\delta y = 5\,(x + \delta x)^{-1} - 5\,x^{-1}$ using the binomial expansion

$$(x + \delta x)^{-1} = x^{-1}\left(1 + \frac{\delta x}{x}\right)^{-1}$$

$$= x^{-1}\left(1 + (-1)\,\frac{\delta x}{x} + (-1)\,(-2)\left(\frac{\delta x}{x}\right)^2 \frac{1}{2!} + \ldots\right)$$

6

$$= x^{-1} \left(1 - \frac{\delta x}{x} + \frac{\delta x^2}{x^2} + \ldots \right)$$

$$\delta y = 5 x^{-1} \left(1 - \frac{\delta x}{x} + \frac{\delta x^2}{x^2} + \ldots \right) - 5 x^{-1} \quad = -5 x^{-1} \frac{\delta x}{x} + 5 x^{-1} \frac{\delta x^2}{x^2} + \ldots$$

dividing each term by δx

$$\frac{\delta y}{\delta x} = -\frac{5 x^{-1}}{x} + 5 x^{-1} \frac{\delta x}{x^2} + \ldots \quad \text{as } \delta x \to 0, \frac{\delta y}{\delta x} \to \frac{dy}{dx} \qquad \frac{dy}{dx} = \frac{-5}{x^2}.$$

(v) $y = 3 x^{1/2}$ ...(1)

$y + \delta y = 3 (x + \delta x)^{1/2}$ ...(2)

(2) - (1)
$\delta y = 3 (x + \delta x)^{1/2} - 3 x^{1/2}$

using the binomial expansion

$$(x + \delta x)^{1/2} = x^{1/2} \left(1 + \frac{\delta x}{x} \right)^{1/2}$$

$$= x^{1/2} \left[1 + \frac{1}{2} \frac{\delta x}{x} + \frac{1}{2} \left(-\frac{1}{2} \right) \left(\frac{\delta x}{x} \right)^2 \frac{1}{2!} + \ldots \right]$$

$$= x^{1/2} + \frac{1}{2} x^{1/2} \frac{\delta x}{x} - \frac{1}{8} \frac{(\delta x)^2}{x^2} x^{1/2} + \ldots$$

$$\delta y = 3 \left[x^{1/2} + \frac{1}{2} x^{1/2} \frac{\delta x}{x} - \frac{1}{8} \left(\frac{\delta x}{x} \right)^2 x^{1/2} + \ldots \right] - 3 x^{1/2}$$

$$\delta y = \frac{3}{2} x^{1/2} \frac{\delta x}{x} - \frac{3}{8} \left(\frac{\delta x}{x} \right)^2 x^{1/2} + \ldots$$

dividing by δx each term

$$\frac{\delta y}{\delta x} = \frac{3}{2} \frac{x^{1/2}}{x} - \frac{3}{8} \frac{\delta x}{x^2} x^{1/2} + \ldots \quad \text{as } \delta x \to 0, \frac{\delta y}{\delta x} \to \frac{dy}{dx}$$

$$\frac{dy}{dx} = \frac{3}{2} \frac{x^{1/2}}{x} = \frac{3}{2 x^{1/2}}$$

WORKED EXAMPLE 2

Determine the derivatives of the functions:-

(i) $\quad y = 3x$ $\qquad$ (ii) $\quad y = 2x^2$ $\qquad$ (iii) $\quad y = 5x^3$

(iv) $\quad y = \dfrac{5}{x}$ $\qquad$ (v) $\quad y = 3x^{1/2}$, using the formula

$$\frac{dy}{dx} = anx^{n-1}, \text{ when } y = ax^n.$$

SOLUTION 2

(i) $\quad y = 3x^1$ $\qquad \dfrac{dy}{dx} = 3 \times 1\,x^{1-1} = 3\,x^0 = 3$

(ii) $\quad y = 2x^2$ $\qquad \dfrac{dy}{dx} = 2 \times 2\,x^{2-1} = 4x$

(iii) $\quad y = 5x^3$ $\qquad \dfrac{dy}{dx} = 5 \times 3\,x^{3-2} = 15\,x^2$

(iv) $\quad y = \dfrac{5}{x} = 5\,x^{-1}$ $\qquad$ expressing in the form $y = ax^n$

$$\frac{dy}{dx} = 5x\,(-1)\,x^{-1-1} = -5\,x^{-2} = -\frac{5}{x^2}$$

(v) $\quad y = 3x^{1/2}$ $\qquad \dfrac{dy}{dx} = 3\left(\dfrac{1}{2}\right)x^{1/2-1} = \dfrac{3}{2}\,x^{-1/2}$

$$= \frac{3}{2\,x^{1/2}}.$$

TO DETERMINE THE DERIVATIVE OF A SUM OR DIFFERENCE OF A FUNCTION

Let $y = f(x) + g(x) - p(x) - q(x)$ ...(1)

where $f(x)$ denotes a function of x and $g(x)$ p, (x), and $q(x)$ are different funtions of x, the derivative of the function (1) is

$$\frac{dy}{dx} = \frac{d}{dx} f(x) + \frac{d}{dx} g(x) - \frac{d}{dx} p(x) - \frac{d}{dx} q(x) \qquad \text{or simply}$$

$$\frac{dy}{dx} = f'(x) + g'(x) - p'(x) - q'(x)$$

where $f'(x)$ denotes the first derivative of $f(x)$ and similarly $g'(x)$, $p'(x)$ and $q'(x)$ denote the first derivatives of the functions $g(x)$, $p(x)$, and $q(x)$ respectively. The differentiation across addition and subtraction is <u>distributive</u>.

WORKED EXAMPLE 3

Determine the derivatives of the functions:

(i) $\qquad y = 3x - 3x^2 - 5x^3$ (ii) $\qquad y = \dfrac{2}{x} - 3x + 4x^2.$

SOLUTION 3

(i) $\qquad y = 3x - 3x^2 - 5x^3$

$$\frac{dy}{dx} = \frac{d}{dx}(3x) - \frac{d}{dx}(3x^2) - \frac{d}{dx}(5x^3)$$

$$= 3 - 6x - 15x^2$$

(ii) $\qquad y = \dfrac{2}{x} - 3x + 4x^2$

$$y = 2x^{-1} - 3x + 4x^2$$

$$\frac{dy}{dx} = -2x^{-2} - 3 + 8x$$

$$= -\frac{2}{x^2} - 3 + 8x.$$

WORKED EXAMPLE 4

Determine the gradients of the following functions at the points adjacent to each function

(i) $\quad y = -2x^{1/2} \quad (x = 1)$ 　　　　　　(ii) $\quad y = 3x^{1/3} \qquad (x = -1)$

(iii) $\quad y = \dfrac{1}{\sqrt{x}} \quad \left(x = \dfrac{1}{2}\right)$ 　　　(iv) $\quad y = \sqrt[3]{x^2} \qquad (x = 1)$

(v) $\quad y = x^2 - x^3 + x^4 \ (x = 0)$ 　　　(vi) $\quad y = \dfrac{1}{x} - \dfrac{1}{x^2} + \dfrac{1}{x^3} \quad (x = 1)$

SOLUTION 4

(i) $\quad y = -2x^{1/2} \qquad \dfrac{dy}{dx} = -x^{-1/2} = -\dfrac{1}{x^{1/2}}$

$\quad$ at $x = 1 \qquad \dfrac{dy}{dx} = -\dfrac{1}{1} = -1$

(ii) $\quad y = 3x^{1/3} \qquad \dfrac{dy}{dx} = x^{-2/3} = \dfrac{1}{x^{2/3}}$

$\quad$ at $x = 1 \qquad \dfrac{dy}{dx} = \dfrac{1}{\sqrt[3]{x^2}} = \dfrac{1}{\sqrt[3]{(-1)^2}} = 1$

(iii) $\quad y = \dfrac{1}{\sqrt{x}} = x^{-1/2} \qquad \dfrac{dy}{dx} = -\dfrac{1}{2}x^{-3/2} = -\dfrac{1}{2\sqrt{x^3}}$

$\quad$ at $x = \dfrac{1}{2} \qquad \dfrac{dy}{dx} = -\dfrac{1}{2\sqrt{1/8}} = -\dfrac{1}{\sqrt{4/8}} = -\sqrt{2}.$

(iv) $\quad y = \sqrt[3]{x^2}$

$\quad y = x^{2/3} \qquad \dfrac{dy}{dx} = \dfrac{2}{3}x^{-1/3}$

10

when $x = 1$ $\dfrac{dy}{dx} = \dfrac{2}{3}$

(v) $y = x^2 - x^3 + x^4$ $\dfrac{dy}{dx} = 2x - 3x^2 + 4x^3$

when $x = 0$, $\dfrac{dy}{dx} = 0.$

(vi) $y = \dfrac{1}{x} - \dfrac{1}{x^2} + \dfrac{1}{x^3} = x^{-1} - x^{-2} + x^{-3}$

$\dfrac{dy}{dx} = -x^{-2} + 2x^{-3} - 3x^{-4} = -\dfrac{1}{x^2} + \dfrac{2}{x^3} - \dfrac{3}{x^4}$

when $x = 1, \dfrac{dy}{dx} = -\dfrac{1}{1} + \dfrac{2}{1} - \dfrac{3}{1} = -2.$

THE DERIVATIVE OF A PRODUCT OF A FUNCTION

PRODUCT RULE

Let $u = f(x)$ and $v = g(x)$

$$y = u.\,v$$

the derivative of a product

$$\boxed{\dfrac{dy}{dx} = v\,\dfrac{du}{dx} + u\,\dfrac{dv}{dx}}$$

WORKED EXAMPLE 5

Determine the derivatives of the following products using the product rule.

(i) $y = x(x^2 + 1)$ (ii) $y = x^{1/2}\left(\sqrt{x} + 1\right)$

(iii) $y = (x + 1)(x^3 - 1)$ (iv) $y = (3x + 5)(5x^2 - 7)$.

SOLUTION 5

(i) $y = x(x^2 + 1)$ where $u = x, v = x^2 + 1$

$$\frac{du}{dx} = 1, \frac{dv}{dx} = 2x$$

$$\frac{dy}{dx} = (x^2 + 1)1 + x\,2x = x^2 + 1 + 2x^2 = 3x^2 + 1$$

(ii) $y = x^{1/2}\left(\sqrt{x} + 1\right)$ where $u = x^{1/2}$ $\dfrac{du}{dx} = \dfrac{1}{2}x^{-1/2}$

$$v = \sqrt{x} + 1 \qquad = x^{1/2} + 1$$

$$\frac{dv}{dx} = \frac{1}{2}x^{-1/2}$$

$$\frac{dy}{dx} = \left(x^{1/2} + 1\right)\frac{1}{2}x^{-1/2} + x^{1/2}\frac{1}{2}x^{-1/2}$$

$$= \frac{1}{2} + \frac{1}{2}x^{-1/2} + \frac{1}{2} = 1 + \frac{1}{2}x^{-1/2}$$

$$= 1 + \frac{1}{2x^{1/2}}$$

(iii) $y = (x + 1).(x^3 - 1)$ where $u = x + 1, \dfrac{du}{dx} = 1$

$$v = x^3 - 1 \qquad \frac{dv}{dx} = 3x^2$$
$$\frac{dy}{dx} = (x^3 - 1)1 + (x + 1)3x^2$$

$$= x^3 - 1 + 3x^3 + 3x^2$$

$$= 4x^3 + 3x^2 - 1$$

(iv) $y = (3x + 5) (5 x^2 - 7)$ where $u = 3x + 5, \dfrac{du}{dx} = 3$

and $v = 5 x^2 - 7, \dfrac{dv}{dx} = 10x$

$\dfrac{dy}{dx} = (5 x^2 - 7) 3 + (3x + 5) 10x$

$= 15 x^2 - 21 + 30 x^2 + 50 x$

$= 45 x^2 + 50 x - 21.$

WORKED EXAMPLE 6

Determine the derivatives of the following functions, using the rule

$y = ax^n, \qquad \dfrac{dy}{dx} = an\, x^{n-1}.$

(i) $y = x (x^2 + 1)$

(ii) $y = x^{1/2} \left(\sqrt{x} + 1 \right)$

(iii) $y = (x + 1) (x^3 - 1)$

(iv) $y = (3 x + 5) (5 x^2 - 7).$

SOLUTION 6

(i) $y = x (x^2 + 1) = x^3 + x, \dfrac{dy}{dx} = 3 x^2 + 1$

(ii) $y = x^{1/2} \left(x^{1/2} + 1 \right) = x + x^{1/2}$ $\dfrac{dy}{dx} = 1 + \dfrac{1}{2} x^{-1/2} = 1 + \dfrac{1}{2 x^{1/2}}$

(iii) $y = (x + 1) (x^3 - 1) = x^4 + x^3 - x - 1$ $\dfrac{dy}{dx} = 4 x^3 + 3 x^2 - 1$

(iv) $y = (3x + 5) (5 x^2 - 7) = 15 x^3 + 25 x^2 - 21x - 35$

$\dfrac{dy}{dx} = 45 x^2 + 50 x - 21$

Observe that the results are the same as in example 5.

THE QUOTIENT RULE

$$y = \frac{u}{v}$$

where $u = f(x)$ and $v = g(x)$ the quotient rule is given by

$$\frac{dy}{dx} = \frac{v \dfrac{du}{dx} - u \dfrac{dv}{dx}}{v^2}$$

WORKED EXAMPLE 7

Determine the derivatives of the quotients (i) $\dfrac{x}{x^2 + 1}$ (ii) $y = \dfrac{3x - 1}{5x^2 - 3}$.

SOLUTION 7

(i) $y = \dfrac{x}{x^2 + 1}$ where $u = x$

and $v = x^2 + 1$, $\dfrac{du}{dx} = 1$ and $\dfrac{dv}{dx} = 2x$

$$\frac{dy}{dx} = \frac{(x^2 + 1).1 - x.2x}{(x^2 + 1)^2}$$

$$= \frac{x^2 + 1 - 2x^2}{(x^2 + 1)^2} = \frac{1 - x^2}{(x^2 + 1)^2}$$

the result should be always simplified.

(ii) $y = \dfrac{3x - 1}{5x^2 - 3}$ where $u = 3x - 1$ and $v = 5x^2 - 3$, $\dfrac{du}{dx} = 3$ and $\dfrac{dv}{dx} = 10$

14

$$\frac{dy}{dx} = \frac{(5x^2 - 3).3 - (3x - 1).10x}{(5x^2 - 3)^2}$$

$$= \frac{15x^2 - 9 - 30x^2 + 10x}{(5x^2 - 3)^2} = \frac{-15x^2 + 10x - 9}{(5x^2 - 3)^2}$$

in simplified form.

WORKED EXAMPLE 8

Determine the derivatives of the functions:-

(i) $y = \dfrac{x^2 + 3}{x}$ (ii) $y = \dfrac{x^3 - 5}{\sqrt{x}}$ using

        a.   the general rule
        b.   the quotient rule.

SOLUTION 8

a.   (i) $y = \dfrac{x^2 + 3}{x} = x + 3x^{-1}$

              $\dfrac{dy}{dx} = 1 - 3x^{-2} = 1 - \dfrac{3}{x^2}$

    (ii) $y = \dfrac{x^3 - 5}{\sqrt{x}} = (x^3 - 5)x^{-1/2} = x^{5/2} - 5x^{-1/2}$

              $\dfrac{dy}{dx} = \dfrac{5}{2}x^{3/2} + \dfrac{5}{2}x^{-3/2}.$

b.   (i) $y = \dfrac{x^2 + 3}{x}$ where $u = x^2 + 3$, $v = x$

                      $\dfrac{du}{dx} = 2x, \dfrac{dv}{dx} = 1$

$$\frac{dy}{dx} = \frac{x \cdot 2x - (x^2 + 3) \cdot 1}{x^2} = \frac{2x^2 - x^2 - 3}{x^2} = \frac{x^2 - 3}{x^2} = 1 - \frac{3}{x^2}.$$

(ii) $\qquad y = \frac{x^2 - 5}{\sqrt{x}}$ where $u = x^3 - 5, \dfrac{du}{dx} = 3x^2$

$$v = x^{1/2} \qquad \frac{dv}{dx} = \frac{1}{2} x^{-1/2}$$

$$\frac{dy}{dx} = \frac{x^{1/2} \cdot 3x^2 - (x^3 - 5) \dfrac{1}{2} x^{-1/2}}{x}$$

$$= \frac{3x^{5/2} - \dfrac{1}{2} x^{5/2} + \dfrac{5}{2} x^{-1/2}}{x} = \frac{\dfrac{5}{2} x^{5/2} + 5/2 \, x^{-1/2}}{x}$$

$$= \frac{5}{2} x^{3/2} + \frac{5}{2} x^{-3/2}.$$

THE FUNCTION OF A FUNCTION

This is best illustrated by an example.

WORKED EXAMPLE 9

Determine the derivative $y = (3x^2 + 5)^{23}$.

SOLUTION 9

$y = (3x^2 + 5)^{23}$

Let $u = 3x^2 + 5$, $\dfrac{du}{dx} = 6x$

$$y = u^{23}, \quad \frac{dy}{du} = 23\, u^{22}$$

$$\frac{dy}{dx} = \frac{dy}{du} \cdot \frac{du}{dx} = 23\, u^{22} \cdot 6x$$

$$= 23\, (3\, x^2 + 5)^{22}\, 6x$$

$$= 138\, x\, (3\, x^2 + 5)^{22}$$

this method is called function of a function, that is, y is a function of u which is a function of x.

WORKED EXAMPLE 10

Determine the derivatives of the functions:-

(i) $\quad y = \dfrac{x^2 - 1}{(x^2 + 1)^{15}}$ (ii) $\quad y = (3\, x^2 - 1)^{1/2}$ (iii) $\quad y = \sqrt{5x - 3}.$

SOLUTION 10

(i) $\quad y = \dfrac{x^2 - 1}{(x^2 + 1)^{15}}$ let $u = 2 - 1, v = (x^2 + 1)^{15}$

$$\frac{du}{dx} = 2x, v = (x^2 + 1)^{15} = W^{15} \text{ where}$$

$$W = x^2 + 1, \frac{dW}{dx} = 2x, \frac{dv}{dW} = 15\, W^{14} = 15\, (x^2 + 1)^{14}$$

$$dv/dx = 30\, x\, (x^2 + 1)^{14}$$

$$dy/dx = \frac{(x^2 + 1)^{15} \cdot 2x - (x^2 - 1) \cdot 30x\, (x^2 + 1)^{14}}{(x^2 + 1)^{30}}$$

$$= (x^2 + 1)^{14}\, \frac{\left[(x^2 + 1)\, 2x - 30\, x^3 + 30x\right]}{(x^2 + 1)^{30}}$$

$$= \frac{2x^3 + 2x - 30x^3 + 30x}{(x^2 + 1)^{16}} = \frac{-28x^3 + 32x}{(x^2 + 1)^{16}}$$

$$= \frac{4x(8 - 7x^2)}{(x^2 + 1)^{16}}.$$

IMPLICIT FUNCTIONS

We have seen that y is expressed in terms of x explicitly, such as $y = 1/x$. This is expressed also implicity such as $xy = 1$.

WORKED EXAMPLE 11

a.   Find dy/dx for the explicit function $y = 1/x$.

a.   Find dy/dx for the explicit function $xy = 1$.

SOLUTION 11

a. $y = \dfrac{1}{x} = x^{-1} \quad \dfrac{dy}{dx} = -x^{-2} = -\dfrac{1}{x^2}$

b. $xy = 1$

differentiating with respect to x, we have

$$\frac{d}{dx}(xy) = \frac{d}{dx}(1)$$

$$\frac{dx}{dx}y + x\frac{dy}{dx} = 0$$

$$y + x\frac{dy}{dx} = 0$$

$$x\frac{dy}{dx} = -y$$

$$\frac{dy}{dx} = -\frac{y}{x} = -\frac{1/x}{x} = -\frac{1}{x^2}.$$

18

The answers are the same as expected, but the second method is more difficult when we use the function implicitly.

WORKED EXAMPLE 12

Determine dy/dx and dx/dy for the implicit function $\dfrac{x^2}{4} + \dfrac{y^2}{9} = 1$.

SOLUTION 12

$$\frac{x^2}{4} + \frac{y^2}{9} = 1.$$

Differentiating with respect to x

$$\frac{d}{dx}\left(\frac{x^2}{4}\right) + \frac{d}{dx}\left(\frac{y^2}{9}\right) = \frac{d}{dx} \quad (1)$$

$$\frac{2x}{4}\frac{dx}{dx} + \frac{2y}{9}\frac{dy}{dx} = 0$$

$$\frac{x}{2} + \frac{2}{9}y \quad \frac{dy}{dx} = 0, \frac{dy}{dx} = -\frac{x/2}{(2\,y/9)\,y} = -\frac{9x}{4y}$$

$$\frac{dy}{dx} = -\frac{9}{4}\frac{x}{y}, \qquad \text{this is expressed in terms of } x \text{ and } y$$

$$\frac{x^2}{4} + \frac{y^2}{9} = 1$$

differentiating with respect to y

$$\frac{d}{dy}\left(\frac{x^2}{4}\right) + \frac{d}{dy}\left(\frac{y^2}{9}\right) = \frac{d}{dy} \quad (1)$$

$$\frac{2x}{4}\frac{dx}{dy} + \frac{2y}{9}\frac{dy}{dy} = 0$$

$$\frac{1}{2}x\frac{dx}{dy} + \frac{2}{9}y = 0, \qquad \frac{dx}{dy} = -\frac{2\,y/9}{\dfrac{1}{2}\,x/2}$$

$$\frac{dx}{dy} = -\frac{4y}{9x}, \text{ the reciprocal of this agrees with } dy/dx = -9x/4y.$$

WORKED EXAMPLE 13

Differentiate with respect to x the function $xy + y^2 x^2 = 3$. Determine dx/dy and check the answer.

SOLUTION 13

$$xy + y^2 x^2 = 3 \dots(1)$$

differentiating with respect to x, we have

$$1. \; y + x \frac{dy}{dx} + 2y \frac{dy}{dx} x^2 + y^2 \, 2x = 0$$

$$\frac{dy}{dx} (x + 2 x^2 y) = - y^2 \, 2x - y$$

$$\frac{dy}{dx} = - \frac{y (1 + 2 xy)}{x (1 + 2 xy)} = - \frac{y}{x}$$

differentiating equation (1) with respect to y, we have

$$\frac{dx}{dy} \cdot y + x.1 + 2y \, x^2 + y^2 + y^2 x \frac{dx}{dy} = 0$$

$$\frac{dx}{dy} (y + 2x \, y^2) = - (x + 2 x^2 y)$$

$$\frac{dx}{dy} = - \frac{- x (1 + 2 xy)}{y (1 + 9xy)} = - \frac{x}{y}$$

the reciprocal is given $\dfrac{dy}{dx} = - \dfrac{y}{x}.$

EXERCISES 1

1. Write down the derivative of the general algebraic form $y = a x^n$.

2. Differentiate from first principles the following algebraic functions:-

   (i) $y = 3$ (ii) $y = x$ (iii) $y = -x^2 + 1$ (iv) $y = 2x^3$

   (v) $y = 5x - \dfrac{3}{x} + \dfrac{1}{x^2}$.

3. Differentiate the following algebraic functions:

   (i) $y = 3x$ (ii) $y = 5$

   (iii) $y = \dfrac{3}{x}$ (iv) $y = -x^2 - x^3 - x^4$

   (v) $y = \dfrac{3}{\sqrt{x}}$ (vi) $y = \dfrac{1}{x} + \dfrac{4}{x^2} - \dfrac{3}{x^3}$

   (vii) $x = 3t^2 - 5t$ (viii) $\Theta = 3t^2 - 5t$

   (ix) $r = \dfrac{1}{t} + t - t^2$ (x) $Z = 5y^2 - 5y^3 - 7y^5$.

4. Differentiate the following products:-

   (i) $y = (x^2 + 3)(x^2 - 5)$ (ii) $t = (x + 1)(x^3 - 9)$

   (iii) $\Theta = (3t^3 - 5)t^5$ (iv) $y = 3(x^{27} - 3)$

   (v) $y = x(x^2 - 1)(x^3 + 2)$.

5. Differentiate the following quotients:-

   (i) $y = \dfrac{x^2}{x^5 - 1}$ (ii) $Z = \dfrac{t^3 - 1}{t^3 + 1}$

(iii) $\Theta = \dfrac{3r}{r_4 - 1}$ (iv) $y = \dfrac{x^4 - 3x^3 + 2x^2}{5x}$

(v) $y = \dfrac{x - 1}{x + 2}$.

6. Differentiate the following functions:-

(i) $y = \sqrt{3x + 1}$ (ii) $y = \dfrac{1}{\sqrt{x - 1}}$

(iii) $y = (x^3 - 1)^3$ (iv) $y = (x^2 - 1)^3 (x^3 + 1)^4$

(v) $y = (5x^2 - 7)^{1/3}$.

7. Distinguish between implicit and expliciy functions.

8. Differentiate the following:-

(i) $xy = c^2$ (ii) $\dfrac{x^2}{a^2} - \dfrac{y^2}{b^2} = 1$

(iii) $\dfrac{y^2}{a^2} - \dfrac{x^2}{b^2} = 1$ (iv) $x^2 + y^2 = r^2$

(v) $\dfrac{x^2}{a^2} + \dfrac{y^2}{b^2} = 1$ (vi) $xy + y^2 = 5$

(vii) $x^2 + y^2 - 3xy + 5y = 0$ (viii) $x^2 + y^2 + 2gx + 2fy + C = 0$

(ix) $y^2 = 4ax$ (x) $x^2 = -5y$.

(a) with respect to x
(b) with respect to y.

9. Determine the gradients at the points showing adjacent to each function.

(i) $y = 3x^2 - 5x - 7$ $(x = -1)$ (ii) $y = \dfrac{x}{x - 1}$ $(x = 2)$

(iii) $y = x(x^3 - 1)(x^2 + 1)$ $(x = 0)$ (iv) $y = 5/x$ $(x = 5)$

(v) $y = -x^{-2/3}$ $(x = 1)$ (vi) $y^2 = x$ $(x = 4)$

(vii) $x^2 = -4y$ $(x = -2)$ (viii) $y = \dfrac{1}{x} + \dfrac{2}{x^2} + \dfrac{3}{x^3}$

                                                                 $(x = -1)$

(ix) $x^2 + y^2 = 4$ ($(x = 1)$ (x) $xy - x^2 + y^2 - 1 = 0$ $(x = 0)$

10.   Differentiate the following functions:-

   (i) $y = an\ x^{n-1}$ (ii) $y = \dfrac{1}{\sqrt{x}} + \sqrt{x} + \sqrt[3]{x^2}$

   (iii) $y = \sqrt{x^2 - 1}\ \sqrt{x^2 + 1}$ (iv) $y = \dfrac{x^2 - 1}{x + 1}$

   (v) $y = \dfrac{x^3 - 1}{x^3 + 2}$ (vi) $y = (1 - 3x)^{1/5}$

   (vii) $y = x^2\sqrt{x - 1}$ (viii) $xy = x^2 + y^2$

   (ix) $3\,x^2 + 3\,y^2 = x - y - 5$ (x) $3\,x^2 + 5\,y^2 = 25.$

11.   Determine the derivatives of the following functions with respect to x:-

   (i) $(2x + 5)^4$ (vi) $(3x - 2)^{1/2}$

   (ii) $(x - 1)^{-4}$ (vii) $(1 + 7x)^{1/2}$

   (iii) $(1 + 3x)^{1/2}$ (viii) $\dfrac{1}{\sqrt{1 + 4x^2}}$

   (iv) $(1 + 2x + 3x^2)^5$ (ix) $(5x^2 + 7x - 3)^2$

   (v) $(5x - 7)^{1/2}$ (x) $\dfrac{x^3 - 2x^2 - 7x + 1}{x^2 - 1}.$

12.   Determine $\dfrac{dy}{dx}$ from the following equations:-

   (i) $\sqrt{x} + \sqrt{y} = \sqrt{3}$ (iv) $5\,y^2 = 4x$

23

(ii) $2x^2 + 3xy = 5$ (v) $2x + 3y = \sqrt{7}$

(iii) $x^2 + y^2 = 3^2$ (vi) $2x^2 + 3y^2 = 25$.

13.   For the function in 12, find the numerical values of $\dfrac{dy}{dx}$ at $x = 1$.

14.   Differentiate with respect to x:-

(i) $\dfrac{x - 7}{x^2 - x + 5}$ (ii) $\dfrac{1}{1 - x}$ (iii) $\dfrac{\sqrt{x} + 1}{\sqrt{x}}$

(iv) $\dfrac{1}{2x^2 + 3x + 4}$ (v) $\dfrac{x - 1}{\sqrt{x}}$.

15.   Find the following:

(i) $\dfrac{d}{dx}(5x^4 + 5x - 1)$

(ii) $\dfrac{d}{dy}\left(\dfrac{y - 1}{y + 1}\right)$

(iii) $\dfrac{d}{dZ}\left(\dfrac{1}{Z} + \dfrac{1}{Z^2} + \dfrac{1}{Z^3}\right)$

(iv) $\dfrac{d}{dt}\left(\dfrac{t^4 + 1}{2t}\right)$

(v) $\dfrac{d}{du}\{(u^2 + 1)(u^3 + 2)\}$

16.   Find the derived functions of the following:-

(i) $y = \dfrac{1}{\sqrt{x}} - \dfrac{1}{x} + \dfrac{1}{x^2}$

(ii) $y = (x + 1)(x + 2)(x + 3)$

(iii) $y = \dfrac{\sqrt{x} - 1}{\sqrt{x} + 1}$

(iv) $\left(1 - \dfrac{1}{x^2}\right)^5$

(v) $xy + x^2 y^2 + x^3 y^3 = 0$.

17.   If $pv = 100$,   (i)   determine dp/dv when $v = 25$
                          (ii)   determine dv/dp when $v = 2$.

2. TRIGNOMETRIC OR CIRCULAR FUNCTIONS

DETERMINE THE DERIVATIVE OF SIN x FROM FIRST PRINCIPLES

$y = \sin x$, $y + \delta y = \sin (x + \delta x)$ if x is increased to $x + \delta x$ then y is increased to $y + \delta y$ subtracting the two equations

$$\delta y = \sin (x + \delta x) - \sin x$$

$\delta y = \sin x \cos \delta x + \sin \delta x \cos x - \sin x$ as $\delta x \to 0$ $\cos \delta x \to 1$ and $\sin \delta x \to \delta x$

$\delta y = \sin x + \delta x \cos x - \sin x$ $\qquad$ $\delta y = \delta y \cos x$

dividing each side by δx,

$$\dfrac{\delta y}{\delta x} = \cos x \qquad \text{as } \delta x \to 0, \qquad \dfrac{\delta y}{\delta x} \to \dfrac{dy}{dx}$$

$$\boxed{\dfrac{dy}{dx} = \cos x}$$

DETERMINE THE DERIVATIVE OF COS x FROM FIRST PRINCIPLES

$y = \cos x$, $y + \delta y = \cos (x + \delta x)$ subtracting the two equations
$\delta y = \cos (x + \delta x) - \cos x$
$\quad = \cos x \cos \delta y - \sin x \sin \delta x, - \cos x$
as $\delta x \to 0$ $\cos \delta x \to 1$, $\sin \delta x \to \delta x$
$\delta y = \cos x - \sin x \delta x - \cos x$
$\delta y = - \sin x \delta x$
dividing each side by δx

$$\dfrac{\delta y}{\delta x} = - \sin x$$

as $\delta x \to 0$, $\dfrac{\delta y}{\delta x} \to \dfrac{dy}{dx}$

$$\boxed{\dfrac{dy}{dx} = -\sin x}$$

Therefore, the derivative of $\sin x$ and $\cos x$ are respectively $\cos x$ and $-\sin x$.

$$y = \sin x \qquad \dfrac{dy}{dx} = \cos x$$

$$y = \cos x \qquad \dfrac{dy}{dx} = -\sin x.$$

THE DERIVATIVE OF TAN x.

$$y = \tan x = \dfrac{\sin x}{\cos x}$$

$$\dfrac{dy}{dx} = \dfrac{\cos x \cos x - \sin x\,(-\sin x)}{\cos^2 x}$$

$$= \dfrac{\cos^2 x + \sin^2 x}{\cos^2 x}$$

$$y = \tan x \qquad \dfrac{dy}{dx} = \sec^2 x$$

THE DERIVATIVE OF COT x

$$y = \cot x = \cos x / \sin x$$

$$\dfrac{dy}{dx} = \dfrac{-\sin x \sin x - \cos x \cos x}{\sin^2 x}$$

$$\dfrac{dy}{dx} = -\dfrac{\sin^2 x + \cos^2 x}{\sin^2 x}$$

$$\dfrac{dy}{dx} = -\operatorname{cosec}^2 x.$$

$$y = \cot x \qquad \dfrac{dy}{dx} = -\operatorname{cosec}^2 x$$

THE DERIVATIVE OF COSEC x.

$$y = \text{cosec } x = \frac{1}{\sin x}$$

$$\frac{dy}{dx} = -\frac{\cos x}{\sin^2 x} = -\frac{\cos x}{\sin x} \cdot \frac{1}{\sin x}$$

$$\frac{dy}{dx} = -\cot x \text{ cosec } x.$$

$$y = \text{cosec } x \qquad \frac{dy}{dx} = -\cot x \text{ cosec } x$$

THE DERIVATIVE OF SEC x.

$$y = \sec x = \frac{1}{\cos x}$$

$$\frac{dy}{dx} = \frac{\sin x}{\cos^2 x} = \frac{\sin x}{\cos x} \cdot \frac{1}{\cos x} = \tan x \sec x$$

$$y = \sec x \qquad \frac{dy}{dx} = \tan x \sec x.$$

FUNCTION y	DERIVATIVE dy/dx
$\sin x$	$\cos x$
$\cos x$	$-\sin x$
$\tan x$	$\sec^2 x$
$\cot x$	$-\text{cosec}^2 x$
$\text{cosec } x$	$-\cot x \text{ cosec } x$
$\sec x$	$\tan x \sec x$

The table above shows the derivatives of the six basic trigoometric functions which must be learnt and written down.

The derivatives of $\sin kx$, $\cos kx$, $\tan kx$ $\cot kx$, $\text{cosec } kx$, $\sec kx$ using function of a function.

Let $u = kx$ $\qquad \dfrac{du}{dx} = k$

$y = \sin kx = \sin u,$ $\qquad dy/du = \cos u$
$y = \cos kx = \cos u,$ $\qquad dy/du = -\sin u$
$y = \tan kx = \tan u,$ $\qquad dy/du = \sec^2 u$
$y = \cot kx = \cot u,$ $\qquad dy/du = -\csc^2 u$
$y = \csc kx = \csc u,$ $\qquad dy/du = -\cot u \csc 2$
$y = \sec kx = \csc u,$ $\qquad dy/du = \tan u \sec u$

$y = \cos kx$ $\qquad \dfrac{dy}{dx} = k \cos kx$

$y = \sin kx$ $\qquad \dfrac{dy}{dx} = -k \sin kx$

$y = \tan kx$ $\qquad \dfrac{dy}{dx} = k \sec^2 kx$

$y = \cot kx$ $\qquad \dfrac{dy}{dx} = k - \csc^2 kx$

$y = \csc kx$ $\qquad \dfrac{dy}{dx} = -k \cot kx \csc kx$

$y = \sec kx$ $\qquad \dfrac{dy}{dx} = -k \tan kx \sec kx$

WORKED EXAMPLE 14

Differentiate the following functions:-

(i) $\quad y = 3 \sin x$ $\qquad\qquad$ (ii) $\quad y = -\sin 3x$

(iii) $\quad y = 2 \cos x$ $\qquad\qquad$ (iv) $\quad y = -4 \cos \dfrac{1}{2} x$

(v) $\quad y = \tan x$ $\qquad\qquad$ (vi) $\quad y = 3 \tan 3x$

(vii) $\quad y = -\cot x$ $\qquad\qquad$ (viii) $\quad y = \cot 5x$

(ix) $\quad y = \csc \dfrac{1}{2} x$ $\qquad\qquad$ (x) $\quad y = 3 \csc 3x$

(xi) $y = \sec\ \dfrac{1}{3}x$ (xii) $y = 3\sec 4x.$

SOLUTION 14

(i) $y = 3\sin x$ $\dfrac{dy}{dx} = 3\cos x$

(ii) $y = -\sin 3x$ $\dfrac{dy}{dx} = -3\cos 3x$

(iii) $y = 2\cos x$ $\dfrac{dy}{dx} = -2\sin x$

(iv) $y = -4\cos\ \dfrac{1}{2}x$ $\dfrac{dy}{dx} = 2\sin\ \dfrac{1}{2}x$

(v) $y = \tan x$ $\dfrac{dy}{dx} = \sec^2 x$

(vi) $y = 3\tan 3x$ $\dfrac{dy}{dx} = 9\sec^2 3x$

(vii) $y = -\cot x$ $\dfrac{dy}{dx} = \operatorname{cosec}^2 x$

(viii) $y = \cot 5x$ $\dfrac{dy}{dx} = -5\operatorname{cosec}^2 5x$

(ix) $y = \operatorname{cosec}\ \dfrac{1}{2}x$ $\dfrac{dy}{dx} = -\dfrac{1}{2}\operatorname{cosec}\ \dfrac{1}{2}x \cot \dfrac{1}{2}x$

(x) $y = 3\operatorname{cosec} 3x$ $\dfrac{dy}{dx} = -9\operatorname{cosec} 3x \cot 3x$

(xi) $y = \sec\ \dfrac{1}{3}x$ $\dfrac{dy}{dx} = \dfrac{1}{3}\sec\ \dfrac{1}{3}x \tan \dfrac{1}{3}x$

(xii) $y = 3\sec 4x$ $\dfrac{dy}{dx} = 12\sec 4x \tan 4x.$

WORKED EXAMPLE 15

Determine the derivatives of the following circular functions:

(i) $y = \sin x \cos x$ (ii) $y = \sin x \tan x$

(iii) $y = \cos x \tan x$ (iv) $y = \cos x \cos x$

(v) $y = \operatorname{cosec} x \sec x.$

SOLUTION 15

(i) $y = \sin x \cos x$

using the product rule

$$\frac{dy}{dx} = \cos x \cos x + \sin x \,(-\sin x)$$

$$= \cos^2 - \sin^2 x = \cos 2x \quad \text{alternatively}$$

$$y = \sin x \cos x = \frac{\sin 2x}{2}$$

$$\frac{dy}{dx} = \frac{2}{2} \cos 2x = \cos 2x$$

(ii) $y = \sin x \tan x$

$$\frac{dy}{dx} = \cos x \tan x + \sin x \sec^2 x$$

$$= \sin x \,(1 + \sec^2 x)$$

(iii) $y = \cos x \tan x = \cos x \,\dfrac{\sin x}{\cos x} = \sin x$

$$\frac{dy}{dx} = \cos x$$

(iv) $y = \cos x \cot x$

$$\frac{dy}{dx} = - \sin x \cot x + \cos x \, (- \text{cosec}^2 \, x)$$

$$= - \cos x \, (1 + \text{cosec}^2 \, x)$$

(v) $y = \text{cosec} \, x \sec x$

$$\frac{dy}{dx} = - \text{cosec} \, x \cot x \sec x + \text{cosec} \, x \tan x \sec x$$

$$\frac{dy}{dx} = - \frac{1}{\sin x} \cdot \frac{\cos x}{\sin x} \cdot \frac{1}{\cos x} + \frac{1}{\sin x} \cdot \frac{\sin x}{\cos x} \cdot \frac{1}{\cos x}$$

$$= - \text{cosec}^2 \, x + \sec^2 \, x.$$

WORKED EXAMPLE 16

Determine the gradients of the following functions at the points indicated aadjacent each function:

(i) $y = 2 \sin x$ $\left(x = \dfrac{\pi}{2} \right)$ (ii) $y = 3 \tan x$ $\left(x = \dfrac{\pi}{4} \right)$

(iii) $y = 3 \cos \dfrac{1}{2} x \ (x = \pi)$ (iv) $y = - \cot x$ $\left(x = \dfrac{\pi}{4} \right)$

(v) $y = \dfrac{\pi}{2} \sin x$ $\left(x = \dfrac{\pi}{4} \right)$ (vi) $y = \text{cosec} \, \dfrac{1}{2} x \ (x = \pi)$

(vii) $y = 3 \sec 3x$ $\left(x = \dfrac{\pi}{6} \right)$ (viii) $y = 5 \cos 5x$ $\left(x = \dfrac{\pi}{2} \right)$

(ix) $y = 2 \sin x \tan x$ $\left(x = \dfrac{\pi}{4} \right)$ (x) $y = 5 \cos x \cot x$ $\left(x = 3 \dfrac{\pi}{2} \right)$

SOLUTION 16

(i) $y = 2 \sin x$ $\dfrac{dy}{dx} = 2 \cos x = 2 \cos \dfrac{\pi}{2} = 0$

(ii) $y = 3 \tan x$ $\dfrac{dy}{dx} = 3 \sec^2 x = 3 \sec^2 \dfrac{\pi}{4} = 6$

(iii) $y = 3 \cos \dfrac{1}{2} x$ $\dfrac{dy}{dx} = -\dfrac{3}{2} \sin \dfrac{1}{2} x = -\dfrac{3}{2}$

(iv) $y = -\cot x$ $\dfrac{dy}{dx} = \operatorname{cosec}^2 x = \operatorname{cosec}^2 \dfrac{\pi}{4} = 2$

(v) $y = \dfrac{\pi}{2} \sin 2x$ $\dfrac{dy}{dx} = \dfrac{\pi}{2} 2 \cos 2x = \pi \cos \dfrac{\pi}{2} = 0$

(vi) $y = \operatorname{cosec} \dfrac{1}{2} x$ $\dfrac{dy}{dx} = -\dfrac{1}{2} \operatorname{cosec} \dfrac{1}{2} x \cot \dfrac{1}{2} x$

$$= -\dfrac{1}{2} \operatorname{cosec} \dfrac{\pi}{2} \cot \dfrac{\pi}{2} = 0$$

(vii) $y = 3 \sec 3x$ $\dfrac{dy}{dx} = 9 \sec 3x \tan 3x$

$$= 9 \sec 3 \dfrac{\pi}{6} \tan 3 \dfrac{\pi}{6}$$

$$= \infty$$

(viii) $y = 5 \cos 5x$ $\dfrac{dy}{dx} = -25 \sin 5x$

$$= -25 \sin 5 \dfrac{\pi}{2} = -25$$

(ix) $y = 2 \sin x \tan x$

$$\dfrac{dy}{dx} = 2 \cos x \tan x + 2 \sin x \sec^2 x$$

$$2 \cos \frac{\pi}{4} \tan \frac{\pi}{4} + 2 \sin \frac{\pi}{4} \sec^2 \frac{\pi}{4}$$

$$= 2 \cdot \frac{1}{\sqrt{2}} \cdot 1 + 2 \cdot \frac{1}{\sqrt{2}} \cdot 2$$

$$= \frac{6}{\sqrt{2}} \frac{\sqrt{2}}{\sqrt{2}} = 3\sqrt{2},$$

(x) $y = 5 \cos x \cot x$

$$\frac{dy}{dx} = -5 \sin x \cot x - 5 \cos x \operatorname{cosec}^2 x$$

$$= -5 \sin \frac{3\pi}{2} \cot \frac{3\pi}{2} - 5 \cos \frac{3\pi}{2} \operatorname{cosec}^2 \frac{3\pi}{2}$$

$$= 0$$

WORKED EXAMPLE 17

Differentiate the following functions:-

(i) $y = 3 \sin (x - \pi/3)$ (ii) $y = -5 \cos \quad (\alpha - x)$

(iii) $y = \cot (x + \pi/2)$ (iv) $y = 5 \tan \quad (3x - 2\pi/5)$

(v) $y = \dfrac{5 \sin x}{\cos (x - \pi/2}$ (vi) $y = \tan 2x \sec \alpha.$

SOLUTION 17

(i) $y = 3 \sin (x - \pi/3)$ $\dfrac{dy}{dx} = 3 \cos (x - \pi/3)$

(ii) $y = -5 \cos (\alpha - x)$ $\dfrac{dy}{dx} = -(-5 \sin (\alpha - x))(-1)$

$$= -5 \sin (\alpha - x)$$

(iii) $y = \cot(x + \pi/2)$ $\dfrac{dy}{dx} = -\,\text{cosec}^2\,(x + \pi/2)$

(iv) $y = 5\tan(3x - 2\pi/5)$ $\dfrac{dy}{dx} = 5\sec^2\,(3x - 2\pi/5)$

(v) $y = \dfrac{5\sin x}{\cos(x - \pi/2)} = \dfrac{5\sin x}{\sin x} = 5$

$\dfrac{dy}{dx} = 0$

(vi) $y = \tan 2x \sec \alpha$

$\dfrac{dy}{dx} = 2\sec \alpha \sec^2 2x$

WORKED EXAMPLE 18

Differentiate the following functions:-

(i) $y = x^2 \sin x$ (ii) $y = \sin^3 x$ (iii) $y = 5\sin^2 x \cos^3 x$

SOLUTION 18

(i) $y = x^2 \sin x$ $\dfrac{dy}{dx} = 2x \sin x + x^2 \cos x$

(ii) $y = \sin^3 x$ Let $u = \sin x$ $\dfrac{dy}{dx} = \cos x$

$y = u^3$ $\dfrac{dy}{du} = 3\,u^2$ $\dfrac{dy}{dx} = 3\cos x \sin^2 x$

(iii) $y = 5\sin^2 x \cos^3 x$

$\dfrac{dy}{dx} = 10 \sin x \cos^4 x + 15 \sin^9 x \,(-\sin x)\cos^2 x$

$$= 10 \sin \quad x \cos^4 x - 15 \sin^3 x \cos^2 x.$$

IMPLICIT FUNCTIONS

If $y \tan x = y^2 x^2 + 3$, determine the value of $\dfrac{dy}{dx}$

differentiating with respect to x $\quad \dfrac{dy}{dx} \tan x + y \sec^2 x = 2y \dfrac{dy}{dx} x^2 + y^2 2x.$

$$\dfrac{dy}{dx} (\tan x - 2y x^2) = 2 xy^2 - y \sec^2 x$$

$$\dfrac{dy}{dx} = \dfrac{y (2xy - \sec x)}{\tan x - 2y x^2}$$

FUNCTION OF A FUNCTION OF A FUNCTION

$$y = 5 \sin^3 (x^2 + 1)$$

Let $u = x^2 + 1$ $\qquad\qquad \dfrac{du}{dx} = 2x$

$$y = 5 \sin^3 u$$

Let $w = \sin u$ $\qquad\qquad \dfrac{dw}{du} = \cos u$

$y = 5 w^3$ $\qquad \dfrac{dy}{dx} 15 w^2$

$$\dfrac{dy}{dx} = \dfrac{dy}{dw} \cdot \dfrac{dw}{du} \cdot \dfrac{du}{dx} = 15 w^2. \cos u. 2x$$

$$= 30 x \sin^2 (x^2 + 1) \cos (x^2 + 1)$$

INVERSE TRIGONOMETRIC FUNCTIONS

$y = \sin^{-1} x$ $\qquad\qquad\qquad y = \cos^{-1} x$

$x = \sin x$

$$\frac{dx}{dy} = \cos y$$

$$\frac{dy}{dx} = \frac{1}{\sqrt{1 - \sin^2 y}}$$

$$\frac{dy}{dx} = \frac{1}{\sqrt{1 - x^2}}$$

$y = \tan^{-1} x$

$x = \tan y$

$$\frac{dx}{dy} = \sec^2 y$$

$$\frac{dy}{dx} = \frac{1}{1 + \tan^2 y}$$

$$\frac{dy}{dx} = \frac{1}{1 + x^2}$$

$y = \sec^{-1} x$

$x = \sec y$

$$\frac{dx}{dy} = \sec y \tan y$$

$$\frac{dy}{dx} = \frac{1}{\sec y \tan y}$$

$$\frac{dy}{dx} = \frac{1}{x (\sec^2 - 1)^{1/2}}$$

$$\frac{dy}{dx} = \frac{1}{x (x^2 - 1)^{1/2}}$$

$x = \cos y$

$$\frac{dx}{dy} = - \sin y$$

$$\frac{dy}{dx} = - \frac{1}{\sqrt{1 - \cos^2 y}}$$

$$\frac{dy}{dx} = - \frac{1}{\sqrt{1 - x^2}}$$

$y = \cot^{-1} x$

$x = \cot y$

$$\frac{dx}{dy} = - \csc^2 y$$

$$\frac{dy}{dx} = - \frac{1}{1 + \cot^2 y}$$

$$\frac{dy}{dx} = - \frac{1}{1 + x^2}$$

$y = \csc^{-1} x$

$x = \csc y$

$$\frac{dx}{dy} = - \csc y \cot y$$

$$\frac{dy}{dx} = - \frac{1}{\csc y \cot y}$$

$$\frac{dy}{dx} = - \frac{1}{x (\cos^2 y - 1)^{1/2}}$$

$$\frac{dy}{dx} = - \frac{1}{x (x^2 - 1)^{1/2}}$$

identities

$$\sin^2 x + \cos^2 y = 1$$
$$1 + \tan^2 y = \sec^2 y$$
$$1 + \cot^2 y = \csc^2 y$$

FUNCTION y	DERIVATIVE dy/dx
$\sin^{-1} x$	$\dfrac{1}{(1 - x^2)^{1/2}}$
$\cos^{-1} x$	$-\dfrac{1}{(1 - x^2)^{1/2}}$
$\tan^{-1} x$	$\dfrac{1}{1 + x^2}$
$\cot^{-1} x$	$-\dfrac{1}{1 + x^2}$
$\sec^{-1} x$	$\dfrac{1}{x (x^2 - 1)^{1/2}}$
$\csc^{-1} x$	$-\dfrac{1}{x (x^2 - 1)^{1/2}}$

EXERCISES 2

1. Differentiate from first principles the following circular functions.

   (i) $y = 3 \sin x$

   (ii) $y = -2 \cos x$

   (iii) $y = \tan 2x$.

2. State the derivatives of the basic trigonometric functions.

   (i) $y = \sin x$

   (ii) $y = 2 \cos x$

   (iii) $y = 3 \tan x$

   (iv) $y = 4 \cot x$

   (v) $y = 5 \operatorname{cosec} x$

   (vi) $y = 6 \sec x$.

3. Determine the gradients of the following trigonometric functions:-

   (i) $y = \dfrac{1}{2} \sin 2x$

   (ii) $y = 3 \cos \dfrac{1}{3} x$

   (iii) $y = 4 \tan \dfrac{1}{4} x$

   (iv) $y = \dfrac{1}{5} \operatorname{cosec} 5x$

   (v) $y = 7 \sec 7x$

   (vi) $y = 2 \cot 3x$.

4. Evaluate the gradients of the functions of question 3 at the points:-

   (a) $x = 0$
   (b) $x = \pi/4$
   (c) $x = 3\pi/4$.

5. Differentiate with respect to x:

   (i) $y = x \sin x$

   (ii) $y = x^2 \sin^2 x$

   (iii) $y = \tan^2 x$

   (iv) $y = 3 \sec^2 x \tan x$

   (v) $y = 5 \operatorname{cosec}^3 x$

   (vi) $y = \cot^4 x$.

6. Differentiate with respect to t:

   (i) $y = 2 \sin^{3/4} t$

   (ii) $y = -\tan^{5/2} 3t$.

(ii) $y = \sqrt{\csc t}$.

7.   Differentiate   (i) $x^2 \sqrt{\cos x}$

       (ii) $x \sqrt{\sin x}$

       (iii) $x \sqrt{\tan x}$.

with respect to x.

8.   If $xy = \tan y$, show that $\dfrac{dy}{dx} = \dfrac{y}{1 + x^2 y^2 - x}$

9.   If $y = \dfrac{\sqrt{1 - x^2}}{\cos^{-1} x}$ find $\dfrac{dy}{dx}$.

10.   If $y = \dfrac{\sin^{-1} x}{\sqrt{1 - x^2}}$ find $\dfrac{dy}{dx}$. provided that $-1 < x < 1$.

11.   Differentiate with respect to x $y = \dfrac{\tan^{-1} x}{\cot^{-1} x}$

12.   Find $\dfrac{dy}{dx}$ if $y = \dfrac{\csc^{-1} x}{\sec^{-1} x}$.

13.   If $y = \dfrac{\cos^{-1} x}{\sin^{-1} x}$, find $\dfrac{dy}{dx}$.

14.   Differentiate the following inverse trigonometric functions:-

       (i) $y = 3 \arcsin 3x$
       (ii) $y = -\arctan 2x$
       (iii) $y = 5 \arccos 4x$.

NOTE: $\arcsin x = \sin^{-1} x$, $\arctan x = \tan^{-1} x$ and $\arccos x = \cos^{-1} x$.

15.   Differentiate the following:-

       (i) $y = \sin x \cos^{-1} x$
       (ii) $y = 2 \cos x \sin^{-1} x$
       (iii) $y = 3 \tan x \cot^{-1} x$.

3. EXPONENTIAL FUNCTIONS

$$y = a^x$$

where the exponent x is the independent variable a is a constant and y is the dependent variable.

If $a = 2$

$$y = 2^x$$

x	0	1	2	3	- 1	- 2	- 3
y	2^0	2^1	2^2	2^3	2^{-1}	2^{-1}	2^{-3}
y	1	2	4	9	$\frac{1}{2}$	$\frac{1}{4}$	$\frac{1}{8}$

when x varies between - 3 and + 3 the curve looks like **Fig. 1**

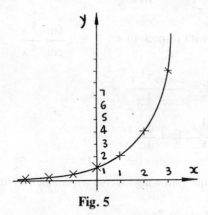

Fig. 5

y increases rapidly, abruptly or exponentially.

If $a = 3$

$$y = 3^x$$

40

x	- 3	- 2	- 1	0	1	2	3
y	$\dfrac{1}{27}$	$\dfrac{1}{8}$	$\dfrac{1}{3}$	1	3	9	27

Plotting the values of y against x we have also an exponential graph as shown in

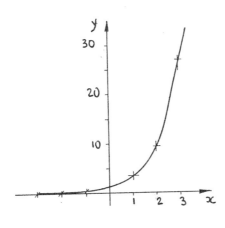

Fig. 6

y increases more abruptly than the previous expression $y = 2^x$.

There is a value between 2 and 3 such that the gradient of the function at certain point is the same as the function, this is the only function in Mathematics whose gradient is the same as the function, $2 < e < 3$

$$y = e^x$$

$$e^x = 1 + \frac{x}{1!} + \frac{x^2}{2!} + \frac{x^3}{3!} + \frac{x^4}{4!} + \dots \qquad (1)$$

e^x is expressed as an infinite e algebraic power series as shown above.

$$e^1 = 1 + \frac{1}{1!} + \frac{1^2}{2!} + \frac{1^3}{3!} + \dots$$

$e = 2.718281828$ to ten significant figures.

Differentiating equation (1) with respect to x, we have

$$\frac{d}{dx}(e^x) = \frac{d}{dx}\left(1 + \frac{x}{1!} + \frac{x^2}{2!} + \frac{x^3}{3!} + \frac{x^4}{4!} + \ldots\right)$$

$$\frac{dy}{dx} = \frac{1}{1!} + \frac{2x}{2!} + \frac{3x^2}{3!} + \frac{4x^3}{4!} + \ldots$$

$$= 1 + \frac{x}{1!} + \frac{x^2}{2!} + \frac{x^3}{3!} + \ldots$$

$4! = 1 \times 2 \times 3 \times 4$ (factorial 4)

therefore

$$\boxed{y = e^x} \quad , \quad \boxed{\frac{dy}{dx} = e^x}$$

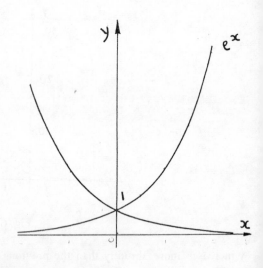

Fig. 7

DETERMINE THE DERIVATIVE OF $y = e^x$ FROM FIRST PRINCIPLES

$$y = e^x, \qquad\qquad y + \delta y = e^{x + \delta x}$$

substracting the equations

$$\delta y = e^{x + \delta x} - e^x$$
$$= e^x . e^{\delta x} - e^x$$
$$= e^x (e^{\delta x} - 1)$$

$$= e^x \left(1 + \frac{\delta x}{1!} + \frac{\delta x^2}{2!} + \frac{\delta x^3}{3!} + \ldots - 1\right)$$

$$\delta y = e^x \left(\frac{\delta x}{1!} + \frac{\delta x^2}{2!} + \frac{\delta x^3}{3!} + \ldots \right)$$

dividing each side by δx

$$\frac{\delta y}{\delta x} = e^x \left(\frac{1}{1!} + \frac{\delta x}{2!} + \frac{\delta x^2}{3!} + \ldots \right)$$

as $\delta x \to 0, \quad \frac{\delta y}{\delta x} \to \frac{dy}{dx}$

$$\boxed{\frac{dy}{dx} = e^x}$$

THE DERIVATIVE OF $y = e^{kx}$.

Let $u = kx$,
$$\frac{du}{dx} = K$$
$$y = e^u, \quad \frac{dy}{du} = e^u$$

$$\frac{dy}{dx} = \frac{dy}{du} \frac{du}{dx} = e^u \cdot k = k \, e^{kx}$$

$$\boxed{\frac{dy}{dx} = k \, e^{kx}}$$

WORKED EXAMPLE 20

Differentiating the following exponential functions:-

(i) $y = 3 \, e^x$ (ii) $y = 3x^{x/3}$ (iii) $y = e^{-3x}$

(iv) $y = 5 \, e^{-5x}$ (v) $y = 2 \, e^{x^2}$ (vi) $y = e^{-x^2}$

(vii) $y = e^{nx}$.

SOLUTION 20

(i) $y = 3 e^x$ $\dfrac{dy}{dx} = 3 e^x$

(ii) $y = 3 \quad e^{x/3}$ $\dfrac{dy}{dx} = 3 \left(\dfrac{1}{3}\right) e^{x/3} = e^{x/3}$

(iii) $y = e^{-3x}$ $\dfrac{dy}{dx} = -3 e^{-3x}$

(iv) $y = 5 e^{-5x}$ $\dfrac{dy}{dx} = -25 e^{-5x}$

(v) $y = 2 \quad e^{x^2}$ let $u = x^2$, $\dfrac{du}{dx} = 2x$

       $y = 2 e^u$ $\dfrac{dy}{du} = 2 e^u \qquad \dfrac{dy}{dx} = 2x . 2 e^{x^2}$

                                                  $\dfrac{dy}{dx} = 4x e^{x^2}$

(vi) $y = e^{-x^2}$ $\dfrac{dy}{dx} = -2x e^{-x^2}$

(vii) $y = e^{nx}$ $\dfrac{dy}{dx} = n e^{nx}.$

WORKED EXAMPLE 21

Differentiate the following functions:-

(i) $y = x e^x$ (ii) $y = e^x \sin x$ (iii) $y = x e^{x^2}$

(iv) $y = x^2 e^{x^2}$ (v) $y = e^{x^2} \cos x$ (vi) $y = e^{-x} \tan x.$

SOLUTION 21

(i) $y = x e^x$ $\dfrac{dy}{dx} = e^x + x e^x$

(ii) $y = e^x \sin x$ $\dfrac{dy}{dx} = e^x \sin x + e^x \cos x$

(iii) $y = x \; e^{x^2}$ $\dfrac{dy}{dx} = e^{x^2} + x \,.\, 2x \, e^{x^2}$

$$= e^{x^2} + 2\,x^2\, e^{x^2}$$

(iv) $y = x^2 \; e^{x^2}$ $\dfrac{dy}{dx} = 2x \,.\, e^{x^2} + x^2 \,.\, 2x \, e^{x^2}$

$$= 2x\, e^{x^2} + 2\,x^3\, e^{x^2}$$

(v) $y = \; e^{x^2} \cos x$ $\dfrac{dy}{dx} = 2x\, e^{x^2} \,.\, \cos x + e^{x^2} \,(- \sin x)$

$$\dfrac{dy}{dx} = 2x\, e^{x^2} \cos x - e^{x^2} \sin x$$

(vi) $y = e^{-x} \tan x$ $\dfrac{dy}{dx} = - \, e^{-x} \tan x + e^{-x} \sec^2 x$

$$\dfrac{dy}{dx} = e^{-x} \,(\sec^2 x - \tan x)$$

EXERCISES 3

1. Write down the power series of (i) e^x (ii) e^{-x} (iii) e^{2x} (iv) e^{-3x} and hence determine the derivative of the functions.

2. Write down the derivatives of the following exponential functions:-

   (i) $y = 3 e^x$ (ii) $y = e^{-3x}$ (iii) $y = e^{x^2}$

   (iv) $y = e^{-3x^2}$ (v) $y = n\, e^{ax}$ (vi) $y = e^{x/2}$

   (vii) $y = \dfrac{1}{2} e^{-x/2}$

3. Differentiate the following products:-

   (i) $y = 3 x^2 e^{x^2}$ (ii) $y = e^x \sin x$ (iii) $y = e^{-3x} \cos x$

   (iv) $y = 3 e^{-x} \sec x$ (v) $y = e^{3x} (x^3 + 3)$.

4. Determine the gradients of the quotients:-

   (i) $y = \dfrac{e^x}{\sin x}$ (ii) $y = \dfrac{\tan x}{e^{2x}}$ (iii) $y = \dfrac{e^{3x} \sin x}{\cot x}$.

5. Differentiate the following functions with respect to x:-

   (i) $y = x^3 e^{3x}$ (ii) $y = e^{-2x} (x^2 - 1)$

   (iii) $y = 3 e^{2x} \cos 2x$ (iv) $y = \dfrac{e^x}{\tan x}$

   (v) $y = \sin (e^{3x})$ (vi) $y = 3 e^{x^2}$

   (vii) $y = e^{x^2} \sin x^2$ (viii) $y = 3 (e^{3x})^5$

   (ix) $y = e^x - e^{-x}$ (x) $y = e^{2x} + e^{-2x}$

   (xi) $y = \dfrac{e^x + e^{-x}}{e^x - e^{-x}}$ (xii) $y = e^{e^{-x}}$

(xiii) $y = (e^{2x} + e^{-2x})$ (xiv) $y = 3\,a^x$

(xv) $y = 2\,(3^x)$.

6.   Differentiate the following functions:-

(i) $y = e^{nx} \sin Kx$ (ii) $y = e^{-mx} \cos nx$

(iii) $y = \dfrac{e^{ax}}{\cos bx}$ (iv) $y = e^{2x} \cos 5x$

(v) $y = e^x \sin 5x$.

7.   Differentiate the following functions:-

(i) $y = e^{-1/t}$ (ii) $y = e^{\sqrt{u}}$

(iii) $y = e^{-\sin x}$.

8.   Differentiate from first principles the following:-

(i) $y = e^{-x}$ (ii) $y = e^{2x}$.

9.   Sketch the graphs:-

(i) $y = 2\,e^x$ (ii) $y = 3\,e^{-x}$.

10.   Differentiating from first principles $y = e^{1/x}$.

4.   LOGARITHMIC FUNCTIONS

$$y \ \log_e x \text{ where } x > 0,$$

by definition of the logarithm

$$e^y = x$$

$$x = e^y$$

$$\frac{dx}{dy} = e^y$$

$$\frac{dy}{dx} = \frac{1}{e^y} = \frac{1}{x}$$

$$\boxed{\frac{dy}{dx} = \frac{1}{x}}$$

The derivative of $\log_e x$ is $\dfrac{1}{x}$. $\log_e x$ can be written as $\ln x$

$$\log_e x = \ln x$$

If $y = \ln x$, $\quad \dfrac{dy}{dx} = \dfrac{1}{x}$.

The graph of $y = \ln x$ is shown in **Fig.**

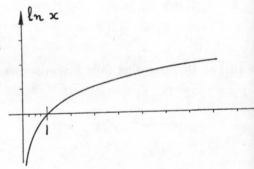

Fig. 8

TO DETERMINE THE DERIVATIVE OF $y = a^x$

$$y = a^x$$

$$\log_e y = x \log_e a$$

$$\frac{d}{dx} (\ln y) = \frac{d}{dx} (x \ln a)$$

$$\frac{1}{y}\frac{dy}{dx} = \ln a$$

$$\frac{dy}{dx} = y \ln a$$

$$\boxed{\frac{dy}{dx} = a^x \ln a}$$

WORKED EXAMPLE 22

Determine the derivatives of the functions:-

(i) $\quad y = 2^x$ (ii) $\quad y = 3^x$ (iii) $\quad y = a^x$

(iv) $\quad y = 5x$ (v) $\quad y = 10^x$.

SOLUTION 22

(i) $\quad y = 2^x$, $\qquad\qquad \frac{dy}{dx} = 2^x \ln 2$

(ii) $\quad y = 3^x$, $\qquad\qquad \frac{dy}{dx} = 3^x \ln 3$

(iii) $\quad y = a^x$, $\qquad\qquad \frac{dy}{dx} = a^x \ln a$

(iv) $\quad y = 5^x$, $\qquad\qquad \frac{dy}{dx} = 5^x \ln 5$

(v) $\quad y = 10^x$, $\qquad\qquad \frac{dy}{dx} = 10^x \ln 10$.

THE DERIVATIVE OF $y = \ln nx$ WHERE n IS A POSITIVE CONSTANT AND $x > 0$.

$$y = \ln nx$$

let $u = nx$, $\quad \dfrac{du}{dx} = n$

$$y = \ln u$$

$$\frac{dy}{du} = \frac{1}{u}$$

$$\frac{dy}{dx} = \frac{dy}{du}\frac{du}{dx} = \frac{1}{u} \cdot n = \frac{1}{nx} \cdot n = \frac{1}{x}$$

$$\boxed{\frac{d}{dx}(\ln nx) = \frac{1}{x}.}$$

Alternatively,

$$y = \ln nx = \ln n + \ln x$$

$$\frac{dy}{dx} = 0 + \frac{1}{x} = \frac{1}{x}$$

$$\frac{d}{dx}(\ln nx) = \frac{1}{x}.$$

Referring to **Fig. 8**, it is observed that the gradient is always positive since $x > 0$.

WORKED EXAMPLE 23

Determine the derivatives of the following logarithmic functions:-

(i) $\quad y = 5 \log_e x$, (ii) $\quad y = \log_{10} x$ (iii) $\quad y = \log_e 3x$

(iv) $\quad y = x \log_e x^2$ (v) $\quad y = x^2 \ln x$ (vi) $\quad y = x \ln x - x$.

SOLUTIONS 23

(i) $y = 5 \log_e x,$ $\dfrac{dy}{dx} = \dfrac{5}{x}$

(ii) $y = \log_{10} x = \log_e x / \log_e 10$

$$\frac{dy}{dx} = \frac{1}{\ln 10} \cdot \frac{1}{x}$$

(iii) $y = \ln 3x, \quad \dfrac{dy}{dx} = 3 \cdot \dfrac{1}{3x} = \dfrac{1}{x}$

(iv) $y = x \log_e x^2$ $\dfrac{dy}{dx} = \ln x^2 + x \cdot \dfrac{2x}{x^2}$

        using the product rule

$$\frac{dy}{dx} = 2 \ln x + 2$$

(v) $y = x^2 \ln x$

$$\frac{dy}{dx} = 2x \cdot \ln x + x^2 \cdot \frac{1}{x} = 2x \ln x + x$$

(vi) $y = x \ln x - x$

$$\frac{dy}{dx} = 1 \cdot \ln x + x \cdot \frac{1}{x} - 1 = \ln x$$

$$\frac{dy}{dx} = \ln x.$$

EXERCISES 4

1. Differentiate with respect to x the following:-

   (i) $\quad y = \log |x|$ $\quad$ (ii) $\quad y = 3 \ln |x|$ $\quad$ (iii) $\quad y = (3x)^x$

   (iv) $\quad y = 7^x$ $\quad\quad\quad$ (v) $\quad y = \ln |Kx|$.

2. Differentiate the following functions with respect to x:

   (i) $\quad y = (\cos x)^x$ $\qquad\qquad\qquad$ (ii) $\quad y = (\cot x)^x$

   (iii) $\quad y = (x + 1)^x$ $\qquad\qquad$ (iv) $\quad y = \sqrt{\dfrac{(x^2 + 1)}{(x^3 - 1)(x^4 + 1.}}$

   (v) $\quad y = \sqrt[3]{\dfrac{(x - 1)}{(x + 1)(x + 2)}}$.

3. Determine the gradients of the functions:-

   (i) $\quad y = \ln 5x^{1/5}$ $\quad$ (ii) $\quad y = \ln \left| \dfrac{1 - x}{x} \right|$ $\quad$ (iii) $\quad y = x^2 \ln x$.

4. Find dy/dx for the following:-

   (i) $\quad y = x^{-3} \ln 3x$ $\quad$ (ii) $\quad y = x - \ln x$ $\quad$ (iii) $\quad y = \dfrac{x}{\ln x}$

   (iv) $\quad y = \dfrac{\ln 2x}{\sin 2x}$ $\quad$ (v) $\quad y = \dfrac{\tan x}{\ln \left| \dfrac{1}{x} \right|}$.

5. If $y = e^x \ln x$ determine dy/dx.

6. If $y = e^{\sin x} \cos (\ln x)$ determine dy/dx.

7. If $y = e^{\cos x} \ln (\sin x)$ determine dy/dx.

8. Differentiate $y = \log_e \sec^2 (5x - 1)$.

9. Determine the first derivatives of the following:-

(i) $y = \sqrt{x(x - 1)(x + 2)}$ (ii) $y = \sqrt{(x + 1)(x + 3)(x + 4)}$.

10. If $y = \sqrt{\dfrac{3x^3+4}{5x^3+7}}$, find dy/dx.

5. HYPERBOLIC FUNCTIONS

$$y = \sinh x = \frac{e^x - e^{-x}}{2}$$

$$\frac{dy}{dx} = \frac{e^x + e^{-x}}{2} = \cosh x$$

$$y = \cosh x = \frac{e^x + e^{-x}}{2}$$

$$\frac{dy}{dx} = \frac{e^x - e^{-x}}{2} = \sinh x$$

$$y = \sinh x \qquad \frac{dy}{dx} = \cosh x$$

$$y = \cosh x \qquad \frac{dy}{dx} = \sinh x$$

$$y = \tanh x = \frac{\sinh x}{\cosh x}$$

$$\frac{dy}{dx} = \frac{\cosh^2 x - \sinh x \sinh x}{\cosh^2 x}$$

$$= \frac{\cosh^2 x - \sinh^2 x}{\cosh^2 x} = \frac{1}{\cosh^2 x}$$

$$= \text{sech}^2 x \text{ where } \cosh^2 x - \sinh^2 x = 1$$

$$y = \tanh x \qquad \frac{dy}{dx} = \text{sech}^2 x$$

$$y = \text{sech } x = \frac{1}{\cosh x}$$

$$\frac{dy}{dx} = \frac{0 . \cosh x - 1 . \sinh x}{\cosh^2 x}$$

$$= - \frac{\sinh x}{\cosh^2 x} = - \frac{\sinh x}{\cosh x} \quad \frac{1}{\cosh x}$$

$$= - \tanh x \text{ sech } x$$

$y = \text{sech } x \qquad \dfrac{dy}{dx} = - \tanh x \text{ sech } x.$

FUNCTION y	DERIVATIVE $\dfrac{dy}{dx}$
sinh x	cosh x
cosh x	sinh x
tanh x	sech^2 x
coth x	- cosech^2 x
cosech x	- coth x cosech x
sech x	- tanh x sech x

WORKED EXAMPLE 24

Write down the derivatives of sinh x, cosh x, tanh x, coth x, cosech x, and sech x.

SOLUTION 24

$\dfrac{d}{dx} (\text{sinh } x) = \cosh x, \dfrac{dy}{dx} (\cosh x) = \sinh x$

$\dfrac{d}{dx} (\tanh x) = \text{sech}^2 x, \dfrac{dy}{dx} (\coth x) = \text{cosech}^2 x$

$\dfrac{d}{dx} (\text{cosech } x) = - \coth x \text{ cosech } x \qquad \dfrac{d}{dx} (\text{sech } x) = - \tanh x \text{ sech } x.$

FUNCTION OF A FUNCTION

$$\frac{d}{dx}(\sinh \ kx) = k \cosh \ kx$$

$$\frac{d}{dx}(\cosh \ kx) = k \sinh \ Kx$$

$$\frac{d}{dx}(\tanh \ kx) = k \ \text{sech}^2 \ kx$$

$$\frac{d}{dx}(\coth \ kx) = k \ \text{cosech}^2 \ kx$$

$$\frac{d}{dx}(\text{cosech} \ kx) = - k \coth \ kx \ \text{cosech} \ kx$$

$$\frac{d}{dx}(\text{sech} \ kx) = - k \tan \ kx \ \text{sech} \ kx.$$

WORKED EXAMPLE 25

Determine the derivatives of the following hyperbolic functions:-

(i) $\qquad y = 3 \sinh 2x$ $\qquad\qquad$ (ii) $\qquad y = - 5 \cosh 3x$

(iii) $\qquad y = \tanh \ \frac{1}{2}x$ $\qquad\qquad$ (iv) $\qquad y = 4 \coth 4x.$

(v) $\qquad y = 2 \ \text{cosech} \ 2x$ $\qquad\qquad$ (vi) $\qquad y = 5 \ \text{sech} \ 5x.$

SOLUTION 25

(i) $\qquad y = 3 \sinh 2x$ $\qquad\qquad\qquad \frac{dy}{dx} = 6 \cosh 2x$

(ii) $\qquad y = - 5 \cosh 3x$ $\qquad\qquad\quad \frac{dy}{dx} = - 15 \sinh 3x$

(iii) $y = \tanh 3x/2$ $\dfrac{dy}{dx} = \dfrac{3}{2} \operatorname{sech}^2 3x/2$

(iv) $y = 4 \coth 4x$ $\dfrac{dy}{dx} = -16 \operatorname{cosech}^2 4x.$

(v) $y = 2 \operatorname{cosech} 2x$ $\dfrac{dy}{dx} = -4 \operatorname{cosech}2x \coth 2x$

(vi) $y = 5 \operatorname{sech} 5x$ $\dfrac{dy}{dx} = 25 \operatorname{sech} 5x \tanh 5x$

WORKED EXAMPLE 26

Determine the derivatives of the following functions:-

(i) $y = \operatorname{sech}^2 x$ (ii) $y = \tanh^3 x$

(iii) $y = \cosh^5 x$ (iv) $y = \sinh^4 x$

(v) $y = \sqrt{2} \operatorname{cosech}^{1/2} x$ (vi) $y = \sqrt{3} \coth^{3/4} x.$

SOLUTION 26

(i) $y = \operatorname{sech}^2$ let $u = \operatorname{sech} x$

     $y = u^2$ $\dfrac{du}{dx} = -\operatorname{sech} x \tanh x$

     $\dfrac{dy}{du} = 2u$ $\dfrac{dy}{dx} = \dfrac{dy}{du} \cdot \dfrac{du}{dx} = -2 \operatorname{sech}^2 \tanh x$

(ii) $y = \tanh^3$ let $u = \tanh x$

     $y = u^3$ $\dfrac{du}{dx} = -\operatorname{sech}^2 x$

     $\dfrac{dy}{du} = 3 u^2$ $\dfrac{dy}{dx} = \dfrac{dy}{du} \cdot \dfrac{du}{dx} = 3 \tanh^2 \operatorname{sech}^2 x$

(iii) $y = \cosh^5 x$ let $u = \cosh x$

$$y = u^5 \qquad \frac{du}{dx} = -\ \text{sech}\ x$$

$$\frac{dy}{du} = 5\ u^4 \qquad \frac{dy}{dx} = 5 \cosh^4 x \sinh x$$

(iv) $y = \sinh^4$ let $u = \sinh x$

$$y = u^4 \qquad \frac{du}{dx} = \cosh x$$

$$\frac{dy}{du} = 4\ u^3 \qquad \frac{dy}{dx} = 4 \sinh^3 x \cosh x$$

(v) $y = \sqrt{2}\ \text{cosech}^{1/2}$, let $u = \text{cosech}\ x$

$$y = \sqrt{2}\ u^{1/2} \qquad\qquad \frac{du}{dx} = \text{cosech}\ x \coth x$$

$$\frac{dy}{du} = \frac{1}{2}\ \sqrt{2}\ u^{-1/2} \qquad \frac{dy}{dx} = -\ \frac{1}{2}\ \sqrt{2}\ \frac{\text{cosech}\ x \coth x}{\text{cosech}^{1/2} x}$$

$$= -\ \frac{1}{2}\ \sqrt{2}\ \text{cosech}^{1/2} x \coth x.$$

(vi) $y = \sqrt{3}\ \coth^{3/4}$, let $u = \coth x$

$$y = \sqrt{3}\ u^{3/4} \qquad\qquad \frac{du}{dx} = -\ \text{cosech}^2 x$$

$$\frac{dy}{du} = \frac{3\sqrt{3}}{4}\ u^{-1/4}, \frac{dy}{dx} = -\ \frac{3\sqrt{3}}{4}\ \frac{\text{cosech}^2 x}{\coth^{1/4} x}.$$

INVERSE HYPERBOLIC FUNCTIONS

$y = \sinh^{-1} x$

$x = \sinh y$

$$\frac{dx}{dy} = \cosh y$$

$$\frac{dy}{dx} = \frac{1}{\cosh y}$$

$$= \frac{1}{\sqrt{1 + \sinh^2 y}}$$

$$= \frac{1}{\sqrt{1 + x^2}}$$

$y = \tanh^{-1} x$

$x = \tanh y$

$$\frac{dx}{dy} = \text{sech}^2 y$$

$$\frac{dy}{dx} = \frac{1}{\text{sech}^2 y}$$

$$= \frac{1}{1 - \tanh^2 y}$$

$$= \frac{1}{1 - x^2}$$

$y = \text{sech}^{-1} x$

$x = \text{sech } y$

$$\frac{dx}{dy} = - \text{sech } y \tanh y$$

$y = \cosh^{-1} x$

$x = \cosh y$

$$\frac{dx}{dy} = \sinh y$$

$$\frac{dy}{dx} = \frac{1}{\sinh y}$$

$$= \frac{1}{\cosh^2 y - 1}$$

$$= \frac{1}{x^2 - 1}$$

$y = \coth^{-1} x$

$x = \coth y$

$$\frac{dx}{dy} = - \text{cosech}^2 y$$

$$\frac{dy}{dx} = \frac{1}{- \text{cosech}^2 y}$$

$$= \frac{1}{1 - \coth^2 y}$$

$$= \frac{1}{1 - x^2}$$

$y = \text{cosech}^{-1} x$

$x = \text{cosech } y$

$$\frac{dy}{dx} = - \coth y \text{ cosech } y$$

$$\frac{dy}{dx} = - \frac{1}{\text{sech } y \text{ tanh } y} \qquad \frac{dy}{dx} = \frac{1}{- \text{coth } y \text{ cosech } y}$$

$$= - \frac{1}{x (1 - \text{sech}^2 y)^{1/2}} \qquad = \frac{1}{- (1 + \text{cosech}^2 y)^{1/2} x}$$

$$= - \frac{1}{x \left(x - x^2 \right)^{1/2}} \qquad = \frac{1}{- \left(1 + x^2 \right)^{1/2} x}$$

Hyperbolic identities

$$\cosh^2 x - \sinh^2 x = 1$$
$$1 - \tanh^2 x = \text{sech}^2 x$$
$$1 - \coth^2 x = - \text{cosech}^2 x$$

FUNCTION y	DERIVATIVE $\dfrac{dy}{dx}$
$\sinh^{-1} x$	$\dfrac{1}{\left(1 + x^2\right)^{1/2}}$
$\cosh^{-1} x$	$\dfrac{1}{\left(x^2 - 1\right)^{1/2}}$
$\tanh^{-1} x$	$\dfrac{1}{1 - x^2}$
$\coth^{-1} x$	$\dfrac{1}{1 - x^2}$
$\operatorname{sech}^{-1} x$	$-\dfrac{1}{x\left(1 - x^2\right)^{1/2}}$
$\operatorname{cosech}^{-1} x$	$-\dfrac{1}{x\left(1 + x^2\right)^{1/2}}$

The graphs of hypebolic functions

The graphs of inverse hyperbolic functions

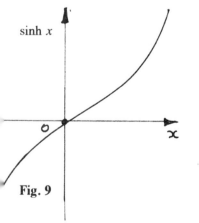

sinh x

Fig. 9

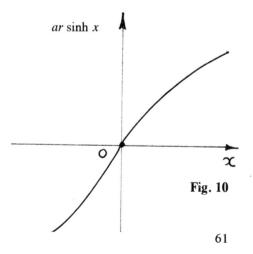

ar sinh x

Fig. 10

61

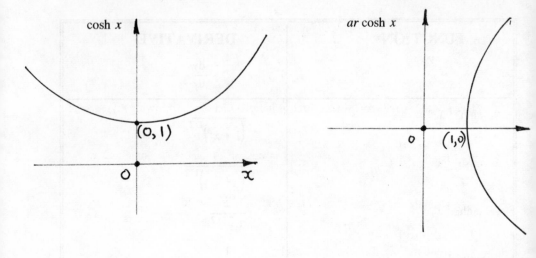

Fig. 11

Fig. 12

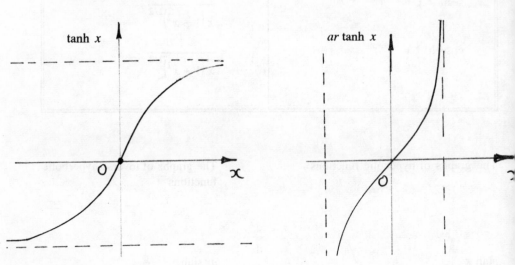

Fig. 13

Fig. 14

EXERCISES 5

1. Derive the derivative from first principles for the following hyperbolic functions:-

    (i) $y = \sinh x$ (ii) $y = \cosh x$ (iii) $\sinh \frac{1}{2} x.$

2. If $y = (\sinh^{-1} 3x)^2$ show that $(1 + 9 x^2) \left(\dfrac{dy}{dx} \right)^2 = 36\, y.$

3. Differentiate the following:-

    (i) $y = \tan 2x \coth 3x$ (ii) $y = \sinh 3x \cot 2x$

    (iii) $y = \operatorname{cosech} \dfrac{1}{x}$ (iv) $y = \operatorname{sech} x^2$

    (v) $y = 3 \sinh^5 \dfrac{x}{2}$ (vi) $y = \coth^{1/2} x \sinh^{3/2} x.$

4. Differentiate the following inverse hyperbolic functions:-

    (i) $y = 3 \sinh^{-1} \dfrac{1}{x}$ (ii) $y = \cosh^{-1} x^2$

    (iii) $y = 5 \cosh^{-1} (x^2 - 3x + 2)$ (iv) $y = \operatorname{cosech}^2 \dfrac{x}{2} \operatorname{sech}^2 \dfrac{x}{3}$

5. If $y = \operatorname{sech}^{-1} 2x$, find $\dfrac{dy}{dx}$ and $\dfrac{d^2y}{dx^2}.$

6. If $y = \sinh^{-1} \dfrac{1}{2} x$, find $\dfrac{d^2y}{dx^2}.$

7. Show that $\dfrac{d}{dx} (\operatorname{cosech}^{-1} 2x) = -\dfrac{1}{x(1 + 4x^2)^{1/2}}.$

8. Show that $\dfrac{d}{dx} (\tanh^{-1} 3x) = -\dfrac{3}{x(1 + 9x^2)}.$

9.   Find the gradients of the functions

(i) $y = \sinh x$ (ii) $y = \cosh x$ (iii) $y = \tanh x$

at (a) $x = -5$ (b) $x = 0$ and (c) $x = 5$, sketch the graphs and indicate these gradients.

10.   Repeat (9) for (i) $y = ar \sinh x$ (ii) $y = ar \cosh x$ (iii) $y = ar \tanh x$ at (a) $x = 2$ (b) $x = 1$.

11.   Differentiate the following functions with respect to x:

(i) $\sin 2x \operatorname{cosech} 3x$ (ii) $\sin 3x$

(iii) $e^x \cosh 2x$ (iv) $\ln \sinh 5x$

(v) $e^{\coth^2 2x}$ (vi) $x^3 \coth^3 5x.$

(vii) $\sqrt{\coth 3x}$ (viii) $\dfrac{1}{3} \cosh^3 x - \cosh x$

(ix) $2 \tanh x \operatorname{sech}^2 x$ (x) $\sqrt{\dfrac{\cosh 2x + 1}{\cosh 2x - 1}}$.

12.   Differentiate:-

(i) $ar \tanh (\cosh 3x)$
(ii) $ar \operatorname{cosech} (\coth 2x)$
(iii) $ar \operatorname{sech} (\tanh x)$
(iv) $ar \tanh (\sinh x)$
(v) $ar \coth (3 x^2 - 1).$

PARAMETRIC EQUATIONS

Certain cartesian functions may be difficult to sketch. If x and y, however, is expressed in terms of a third variable, t, called, <u>the parameter</u>, the functions may be sketched easier.

Consider the cartesian function

$$y^2 = x^2 (x - 1) \ldots (1)$$

whose parametric equations are $x = t^2 + 1$ and $y = t (t^2 \ 1)$.

The lefthand side of (1)

$$y^2 = t^2 (t^2 + 1)^2 = t^2 (t^4 + 2 t^2 + 1)$$

the righthand side of (1)

$$x^2 (x - 1) = (t^2 + 1)^2 (t^2 + 1 - 1)$$

$$= t^2 (t^2 + 1)^2.$$

Therefore the parametric equation of $y^2 = x^2 (x - 1)$ are $x = t^2 + 1$ and $y = t (t^2 + 1)$.

Sketch the cartesian equation $y^2 = x^2 (x - 1)$

if $x = 0, y = 0$; if $x \geq 1, y$ exists

if $x = 1, y = 0$; if $x = 2, y = \pm 2$ and

if $x = 3, y^2 = 9 (2), y = \pm 3 \ \sqrt{2}$ and so on.

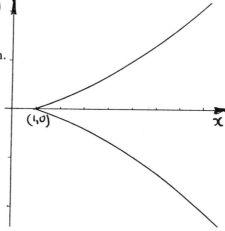

Fig. 15

65

The curve is symmetrical about the x-axis,

Sketch the parametric equations

$$x = t^2 + 1, y = t (t^2 + 1).$$

if $t = 0$; $x = 1$, $y = 0$

if $t = 1$; $x = 2$, $y = 2$

if $t = 2$; $x = 5$, $y = 10$

if $t = -1$ $x = 2$, $y = -2$

if $t = -2$ $x = 5$, $y = -10$

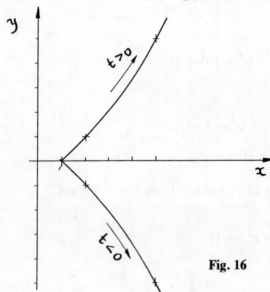

Fig. 16

Differentiating $x = t^2 + 1$ and $y = t (t^2 + 1) = t^3 + t$ with respect to t, we have

$$\frac{dx}{dt} = 2t \quad \text{and} \quad \frac{dy}{dt} = 3 t^2 + 1$$

$$\frac{dy / dx}{dx / dy} = \frac{dy}{dx} = \frac{3 t^2 + 1}{2t}$$

If t is positive dy/dx is positive
If t is negative dy/dx is negative

A curve is given by the parametric equations.

$$x = \Theta - \sin \Theta \ \dots(1)$$

$$y = 1 - \cos \Theta \ \dots(2)$$

Sketch the curve $0 \leq \Theta \leq 2\pi$.

Θ^c	0	$\dfrac{\pi}{6}$	$\dfrac{\pi}{3}$	$\dfrac{\pi}{2}$	$\dfrac{2\pi}{3}$	$\dfrac{5\pi}{6}$
$\sin \Theta^c$	0	0.5	0.866	1	0.866	0.5
$\cos \Theta^c$	1	0.866	0.5	0	- 0.5	- 0.866
$y = 1 - \cos \Theta$	0	0.134	0.5	1	1.5	1.866
$t = \Theta - \sin \Theta$	0	0.024	0.181	0.511	1.228	2.118

cont ...

π	$\dfrac{7\pi}{6}$	$\dfrac{4\pi}{3}$	$\dfrac{3\pi}{2}$	$\dfrac{5\pi}{3}$	$\dfrac{11\pi}{6}$	2π
0	- 0.5	- 0.866	- 1	- 0.866	- 0.5	0
- 1	- 0.866	- 0.5	0	0.5	0.866	1
2	1.866	1.5	1	0.5	0.134	0
3.142	4.165	5.055	5.712	6.102	6.260	6.283

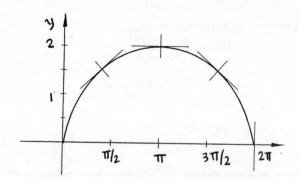

Fig. 17

Determine dy/dx.

Differentiating (1) and (2)

$$\frac{dx}{d\Theta} = 1 - \cos \Theta \qquad \frac{dy}{d\Theta} = \sin \Theta$$

$$\frac{dy}{dx} = \frac{dy/d\Theta}{dx/d\Theta} = \frac{\sin \Theta}{1 - \cos \Theta} = \frac{2 \sin \dfrac{\Theta}{2} \cos \dfrac{\Theta}{2}}{1 - \left(2 \cos^2 \dfrac{\Theta}{2} - 1\right)}$$

$$\frac{dy}{dx} = \frac{2 \sin \dfrac{\Theta}{2} \cos \dfrac{\Theta}{2}}{2 \left(1 - \cos^2 \dfrac{\Theta}{2}\right)} = \frac{2 \sin \dfrac{\Theta}{2} \cos \dfrac{\Theta}{2}}{2 \sin \sin^2 \dfrac{\Theta}{2}} = \cot \frac{\Theta}{2}$$

$$\frac{dy}{dx} = \cot \frac{\Theta}{2}.$$

Determine the values of dy/dx at Θ.

(i) $\qquad \dfrac{\pi}{2}$ (ii) $\quad \pi$ (iii) $\qquad \dfrac{3\pi}{2}$ (iv) $\quad 2\pi.$

(ii) $\qquad \dfrac{dy}{dx} = \cot \dfrac{\Theta}{2} = \cot \dfrac{\pi}{4} = 1$

(iii) $\qquad \dfrac{dy}{dx} = \cot \dfrac{\Theta}{2} = \cot \dfrac{\pi}{2} = 0$

(iv) $\qquad \dfrac{dy}{dx} = \cot \dfrac{\Theta}{2} = \cot \dfrac{3\pi}{2} = -1$

(v) $\qquad \dfrac{dy}{dx} = \cot \dfrac{\Theta}{2} = \cot \pi = -\omega$

Determine the equations of the tangents and normals at $\Theta = \dfrac{\pi}{2}$ and $\Theta = \dfrac{3\pi}{2}$.

At $\Theta = \pi/2 \qquad x = \Theta - \sin \Theta = \dfrac{\pi}{2} - 1$

$\qquad\qquad\qquad\qquad y = 1 - \cos \Theta = 1$

$$y = mx + c = 1.x + c$$

$$1 = \dfrac{\pi}{2} - 1 + c, \; c = 2 - \dfrac{\pi}{2}$$

$$\boxed{y = x + 2 - \dfrac{\pi}{2}}$$

the gradient of the normal is -1 $\quad y = -x + c \quad 1 = -\left(\dfrac{\pi}{2} - 1\right) + c$

$c = \dfrac{\pi}{2}$

$$\boxed{y = -x + \dfrac{\pi}{2}}$$

At $\Theta = \dfrac{3\pi}{2} \quad x = \Theta - \sin \Theta = \dfrac{3\pi}{2} + 1 \quad y = 1 - \cos \Theta = 1$

$y = x + c \qquad 1 = \left(\dfrac{3\pi}{2} + 1\right) + c \qquad\qquad c = -\dfrac{3\pi}{2}$

$$\boxed{y = x - \dfrac{3\pi}{2}}$$

the gradient of the normal is -1

$$y = -x + c \qquad 1 = -\frac{3\pi}{2} - 1 + c \qquad c = 2 + 3 \qquad c = 2 + 3\frac{\pi}{2}$$

$$\boxed{y = -x + 2 + 3\frac{\pi}{2}}$$

The cartesian equation of the cycloid.

Eliminate the parameter between (1) and (2).

$$x = \Theta - \sin\Theta \qquad \sin\Theta = \Theta - x \qquad y = 1 - \cos\Theta \qquad \cos\Theta = 1 - y$$

$$\sin^2\Theta + \cos^2\Theta = 1 \qquad (\Theta - x)^2 + (1 - y)^2 = 1$$

$$\Theta - x = \pm\sqrt{1 - (1-y)^2} = \pm\sqrt{(2-y)y} \qquad \Theta = x \pm\sqrt{y(2-y)}$$

$$1 - y = \cos\left[x \pm\sqrt{y(2-y)}\right] \quad 0 \le y \le 2 \text{ and } 0 \le x \le 2\pi.$$

The cartesian equations is rather complicated.

WORKED EXAMPLE 26

Determine the cartesian equations of the following parametric equations:-

(i) $x = a\cos\Theta$ (ii) $x = at^2$ (iii) $x = ct$

        $y = b\sin\Theta$ $y = 2at$ $y = \frac{c}{t}.$

(iv) $x = r\cos\Theta$ (v) $x = a\cosh\Theta$

        $y = r\sin\Theta$ $y = b\sinh\Theta$

SOLUTION 26

(i) $x = a\cos\Theta, \cos\Theta = x/a \ldots(1)$

      $x = b\sin\Theta, \sin\Theta = y/b \ldots(2)$

      squaring up both sides of (1) and (2) and adding

$$\cos^2 \Theta + \sin^2 \Theta = \left(\frac{x}{a}\right)^2 + \left(\frac{y}{b}\right)^2 = 1$$

$$\frac{x^2}{a^2} + \frac{y^2}{b^2} = 1 \quad \text{an ellipse.}$$

(ii) $x = a\,t^2 \ldots(1)$

$y = 2\,at \ldots(2)$

From (2) $t = y/2a$, substitute in (1) $x = a \left(\frac{y}{2a}\right)^2 = \frac{y^2}{4a}$

$y^2 = 4ax$ a parabola.

(iii) $x = ct \ldots(1)$

$y = c/t \ldots(2)$

Multiplying (1) and (2)

$xy = ct \cdot \dfrac{c}{t} = c^2$

$\quad xy = C^2$ rectangular hyperbola

(iv) $x = r \cos \Theta \ldots(1)$

$y = r \sin \Theta \ldots(2)$

$\cos \Theta = x/r \ldots(3)$

$\sin \Theta = y/r \ldots(4)$

squaring up both sides of (3) and (4) and adding:-

$$\cos^2 \Theta + \sin^2 \Theta = \frac{x^2}{r^2} + \frac{y^2}{r^2} = 1$$

$\quad x^2 + y^2 = r^2$ a circle

(v) $x = a \cosh \Theta, \ldots(1)$ $\cosh \Theta = x/a \ldots(3)$

$y = b \sinh \Theta, \ldots(2)$ $\sinh \Theta = y/b \ldots(4)$

squaring up both sides of (3) and (4) and substracting

$$\cosh^2 \Theta - \sinh^2 \Theta = \frac{x^2}{a^2} - \frac{y^2}{b^2} = 1$$

a hyperbola.

Determine dy/dx and d^2y/dx^2 for parametric equations:-

(i) $x = a \cos \Theta$ $dx/d\Theta = -a \sin \Theta$

$y = b \cos \Theta$ $dx/d\Theta = -b \cos \Theta$

$$\frac{dy}{dx} = \frac{dy/d\Theta}{dx/d\Theta} = -\frac{b}{a} \cot \Theta$$

(ii) $x = a\,t^2,$ $dx/dt = 2at$

$y = 2\,at$ $dy/dt = 2a$

$$\frac{dy}{dx} = \frac{dy/dx}{dx/dt} = \frac{2a}{2at} = \frac{1}{t}$$

(iii) $x = ct$ $dx/dt = c$

$y = c/t = c\,t^{-1},$ $dy/dt = -c/t^2$

$$\frac{dy}{dx} = \frac{dy/dt}{dx/dt} = \frac{-c/t^2}{c/t} = -\frac{1}{t}$$

(iv) $x = r \cos \Theta$ $dx/d\Theta = r \sin \Theta$

$y = r \sin \Theta$ $dx/d\Theta = r \cos \Theta$

$$\frac{dy}{dx} = \frac{dy/d\Theta}{dx/d\Theta} = \frac{r \cos \Theta}{-r \sin \Theta} = -\cot \Theta$$

(v) $x = a \cosh \Theta$ $dx/d\Theta = a \sinh \Theta$

$y = b \sinh \Theta$ $dx/d0 = b \cosh \Theta$

$$\frac{dy}{dx} = \frac{b \cosh \Theta}{a \sinh \Theta} = \frac{b}{a} \coth \Theta.$$

(i)
$$\frac{d^2y}{dx^2} = -\frac{b}{a}\,(-\operatorname{cosec}\Theta\cot\Theta)\,\frac{d\Theta}{dx}$$

$$= \frac{b}{a}\operatorname{cosec}\Theta\cot\Theta\,\frac{1}{-a\sin\Theta}$$

$$= -\frac{b}{a^2}\operatorname{cosec}^2\Theta\cot\Theta$$

(ii)
$$\frac{d^2y}{dx^2} = -\frac{1}{t^2}\cdot\frac{dt}{dx} = -\frac{1}{t^2}\cdot\frac{1}{2at} = -\frac{1}{2at^3}$$

(iii)
$$\frac{d^2y}{dx^2} = -\frac{1}{t^2}\frac{dt}{dx} = \frac{1}{t^2}\cdot\frac{1}{c} = \frac{1}{ct^2}$$

(iv)
$$\frac{d^2y}{dx^2} = \operatorname{cosec}\Theta\cot\Theta.\quad\frac{d\Theta}{dx} = \frac{\operatorname{cosec}\Theta\cot\Theta}{-r\sin\Theta}$$

$$= -\frac{1}{r}\operatorname{cosec}^2\Theta\cot\Theta$$

(v)
$$\frac{d^2y}{dx^2} = -\frac{b}{a}\operatorname{cosech}\Theta\coth\Theta\,\frac{d\Theta}{dx}$$

$$= -\frac{b}{a}\operatorname{cosech}\Theta\coth\Theta\,\frac{1}{a\sinh\Theta}$$

$$= -\frac{b}{a}\operatorname{cosech}^2\Theta\coth\Theta.$$

EXERCISES 6

1. Obtain dy/dx in terms of the parameter t, if $x = 2 \sinh t$, $y = 3 \cosh t$.

2. (a) If $x = t - \sin t$ and $y = 1 - \cos t$ find (i) dy/dx (ii) d^2y/dx^2 in terms of half angles.

   (b) Sketch the curve given by the parametric equations in (a) for $0 \le t \le 2\pi$.

3. If $x = ct$ and $y = c/t$ find (i) dy/dx (ii) d^2y/dx^2 in terms of t. Sketch the curve.

4. Sketch the curve given parametrically by $x = 2\,t^2$ and $y = 2\,t^3$.

5. If $x = 1 + t^2$ and $y = 2t - 1$, find (i) dy/dx (ii) d^2y/dx^2 in terms of t. Sketch the curve.

6. Sketch the curve given parametrically by $x = 4\,t^2$, $y = 4t$. You may consider the gradient dy/dx.

7. If $x = 2 \sin t$ and $y = 2 \cos^3 t$, determine (i) dy/dx and (ii) $\dfrac{d^2y}{dx^2}$ and hence determine the points of inflexions.

   Sketch the curve.

8. A curve is given by the parametric equations
   $x = 2 \cos t + (t + 3) \sin t$, $y = 2 \sin t - (t + 3) \cos t$, $0 \le t \le 2\pi$, $t \ne -3$.

   Determine dy/dx.

9. The parametric equations of a curve are $x = 2 \cos \Theta$ and $y = 4 \sin \Theta$. Find dy/dx.

10. A curve has parametric equations $x = t + e^t$, $y = 2t - e^{2t}$.

    Determine dx/dt dy/dt and hence dy/dx and $\dfrac{d^2y}{dx^2}$.

11. A curve has parametric equations $x = 3t$ and $y = 3 \ln \sec t$. Determine the derivatives:-

    (i) dy/dx and (ii) $\dfrac{d^2y}{dx^2}$.

12. A curve has parametric equations of $x = t - \sin t$, $y = 1 - \cos t$. Find dy/dx.

13. A curve has parametric equations $x = 3 \cos t$ and $y = \cos 2t$. Determine (i) dy/dx and (ii) d^2y/dx^2.

14. A curve is given by the equations $x = 3 \sin 2t (1 - \cos 2t)$, $y = 3 \cos 2t (1 + \cos 2t)$. Find dy/dx.

15. A curve is defined by the parametric equations $x = 2 \cos \Theta - \cos 2\Theta$, $y = 2 \sin \Theta - \sin 2\Theta$. Determine dy/dx and d^2y/dx^2.

16. Find dy/dx at $t = \dfrac{\pi}{2}$ for a curve whose parametric equations are $x = t \sin t - \cos t - 1$, $y = \sin t - t \cos t$.

17. A curve is defined by the parametric equations $x = 3 (\Theta - \sin \Theta)$, $y = 3 (1 - \cos \Theta)$. Show that

$$\frac{d^2y}{dx^2} = - \frac{1}{12} \operatorname{cosec}^4 \frac{\Theta}{2}.$$

18. A curve is given parametrically by the equations:-

$x = t - \tanh t$, $y = \operatorname{sech} t$ find dy/dx and d^2y/dx^2. Hence determine the value of $(dy/dx)^3 - (d^2y/dx^2) + y/x$.

7. SECOND AND HIGHER DERIVATIVES OF A FUNCTION

NOTATION

SECOND DERIVATIVE

$$\frac{d}{dx}\left(\frac{dy}{dx}\right) = \text{the rate of change of}$$

$$\frac{dy}{dx} \quad \text{with respect to } x$$

$$= \frac{d^2y}{dx^2}$$

$$\frac{dy}{dx} = f'(x) \quad \text{when } y = f(x)$$

$$\frac{d^2y}{dx^2} = f''(x) = \text{the second derivative.}$$

THE MEANING OF THE SECOND DERIVATIVE

The first derivative or the primitive of a function denotes the gradient of a function.

$\dfrac{dy}{dx}$ may be positive, $\quad \dfrac{dy}{dx} > 0$

$\dfrac{dy}{dx}$ may be negative, $\quad \dfrac{dy}{dx} > 0$

$\dfrac{dy}{dx}$ may be zero, $\quad \dfrac{dy}{dx} = 0$

To illustrate the various cases above, consider the parabola with a maximum and a minimum.

$$\frac{dy}{dx} = 0$$

$\frac{dy}{dx}$ is changing from a positive to a

negative, the rate of change of $\frac{dy}{dx}$

Fig. 18

$$\frac{dy}{dx} > 0 \qquad\qquad \frac{dy}{dx} < 0 \qquad \text{is} \quad \frac{d^2y}{dx^2} < 0,$$

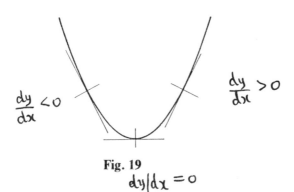

negative, $\frac{d}{dx}\left(\frac{dy}{dx}\right) = \frac{d^2y}{dx^2} < 0,$ therefore, for a maximum the rate of change of

the gradient is negative, at the peak, it is neither positive or negative, it is zero

$$\frac{dy}{dx} = 0$$

for a maximum. Consider a minimum.

$\frac{dy}{dx} < 0$ $\qquad\qquad$ $\frac{dy}{dx} > 0$

Fig. 19

$dy/dx = 0$

the rate of change of $\frac{dy}{dx}$ is positive, $\frac{dy}{dx}$ changes from negative to positive,

via a zero gradient.

WORKED EXAMPLE 27

Skertch the graph $y = \sin x$ and determine the gradients at $x = 0,$

$\frac{\pi}{4}, \frac{\pi}{2}, \frac{3\pi}{4}, \pi, \frac{5\pi}{4}, \frac{3\pi}{3}, \frac{7\pi}{4}$ and $2\pi.$

Ilustrate the primitive or first derivative and second derivatives for the maximum and minimum values of the function.

SOLUTION 27

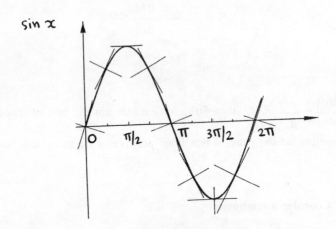

sin *x*

Fig. 20

$y = \sin x \quad \dfrac{dy}{dx} = \cos x \quad$ at $x = 0, \quad \dfrac{dy}{dx} = \cos 0 = 1$

$x = \dfrac{\pi}{4}, \qquad \dfrac{dy}{dx} = \cos \dfrac{\pi}{4} = 0.707 \quad x = \dfrac{\pi}{2}, \qquad \dfrac{dy}{dx} = \cos \dfrac{\pi}{2} = 0$

$x = \dfrac{3\pi}{4}, \qquad \dfrac{dy}{dx} = \cos \dfrac{3\pi}{4} = 0.707 \quad x = \pi \qquad \dfrac{dy}{dx} = \cos \pi = -1$

$x = \dfrac{5\pi}{4}, \qquad \dfrac{dy}{dx} = -0.707 \qquad\qquad x = \dfrac{3\pi}{2}, \qquad \dfrac{dy}{dx} = 0$

$x = \dfrac{7\pi}{4}, \qquad \dfrac{dy}{dx} = 0.707 \qquad\qquad x = 2\pi \quad \dfrac{dy}{dx} = 1.$

For a maximum

at $x = \dfrac{\pi}{4}, \dfrac{dy}{dx} > 0$ at $x = \dfrac{\pi}{2}, \dfrac{dy}{dx} > 0$

at $x = \dfrac{3\pi}{4}, \dfrac{dy}{dx} > 0$ $\dfrac{dy}{dx}\left(\dfrac{dy}{dx}\right) < 0$ or $\dfrac{d^9 y}{dx^2} < 0.$

For a minimum

at $x = \dfrac{5\pi}{4}, \dfrac{dy}{dx} < 0$ at $x = \dfrac{3\pi}{2}, \dfrac{dy}{dx} < 0$ at $x = \dfrac{7\pi}{4}, \dfrac{dy}{dx} < 0$

$\dfrac{dy}{dx}\left(\dfrac{dy}{dx}\right) > 0$ or $\dfrac{d^2 y}{dx^2} > 0.$

EXERCISES 7

1. Determine the first and second derivatives for the following functions:-

   (i) $y = 3x^2 - 5x + 7$
   (ii) $x = t - 6t^2 + 7t^3$
   (iii) $u = 3v^2 + 5v - 1$
   (iv) $w = 3z^2 - z - 4$

2. Determine the second derivatives of the following functions and simplify:-

   (i) $y = \dfrac{3x^2 - 1}{x + 1}$

   (ii) $y = e^x + \sin x$

   (iii) $y = e^{-x}/\cos 2x$

   (iv) $y = 3 \sin 2x - 5 \cos 2x$

   (v) $y = \sin^2 x.$

3. A body is moving along a straight line and its distance x metres from a fixed point on the line after a time t seconds is given by $x = 2t^3 - 3t^2 + 4t + 5$.

   Find

   (i)   the velocity of the body after 1 s,
   (ii)   the velocity of the body at $t = 0$s,
   (iii) the velocity of the body after 5 s from the start
   (iv)   the acceleration at the start and after 2 s
   (v)   the displacements after 2, 3, and 5 s.

4. If the distance S metres a body moves after t s is given by $S = 30t^2 - 3t + 5$.

   Find

   (i)   its velocity after 3 s
   (ii)   its acceleration
   (iii) the distance the body has travelled before coming to rest
   (iv)   the time when the velocity, 57 m/s

(v)   the velocity after 10 s.

5.   A body is falling freely from rest under gravity $(g = 9.81\,\text{m/s}^2)$, the distance S metres travelled is given by the expression $S = 20\,t^2$, where t is the time in seconds.

Find

(i)   the velocity after t seconds,
(ii)   the velocity after 1 second,
(iii)   the time taken for the body to fall 1500 m
(iv)   the acceleration.

6.   Find $\dfrac{d^2y}{dx^2}$ for the following functions:-

(i) $y = 3 \sin 2x - 5 \cos 2x$
(ii) $y = 3 x^3 - 2 x^2 + x - 1$
(iii) $y = 4 e^{-2x} - 5 e^{3x}$
(iv) $y = e^{3x} - \cos 3x + \sin 3x$
(v) $y = 5 \ln x + x \sin 2x.$

7.   Determine the first and second derivatives of the following functions:-

(i) $x = 5 t^5 - 4 t^4 + 3 t^3 - 2 t^2 + t - 1$
(ii) $x = \sin t - \cos t$
(iii) $x = e^t \sin 2t$
(iv) $x = e^{2t}/(1 + t)$
(v) $x = e^{\sin t}.$

8.   Define velocity and acceleration and determine the velocity and acceleration for the functions in exercise 2 at $t = 0$.

9.   Determine the second derivatives for the following functions:-

(i) $y = \dfrac{x^2 + 1}{x - 1}$

(ii) $y = x^2 \sin x$

(iii) $y = \dfrac{\cos x}{e^{3x}}$

(iv) $y = \dfrac{\ln x}{(1 + x)^2}$

(v) $y = e^x/(1 + x)$

10. Evaluate the second derivatives for the functions at $x = \dfrac{\pi}{2}$.

   (i) $y = \sin x \cos 2x$
   (ii) $y = e^x \tan x/2$
   (iii) $y = x^2 \ln x$
   (iv) $y = 5 \cos 3x - 4 \sin 4x$

11. Determine $\dfrac{d^2y}{dx^2}$ for $y = 3 \cos kx$ where k is a count and at $x = \dfrac{\pi}{k}$.

12. What is the significance of $\dfrac{dy}{dx}$ and $\dfrac{d^2y}{dx^2}$ of a function? Illustrate your
     answers clearly with the aid of sketches.

13. Find the second derivative for the function

$$y = e^x \sin 2x.$$

14. Find the second derivatives of the function

$$y = e^{-2x} \cos x.$$

15. A particle moves s m in time t seconds given by the relation
     $s = 3t^3 - t^2 + t + 7$. Find the velocity and the acceleration of the particle
     after 5 seconds.

16. Find the $\dfrac{d^2y}{dx^2}$ of the following quotients:-

   (i) $\dfrac{1 + \cos \Theta}{\sin \Theta}$ (ii) $\dfrac{\sin x}{e^2 x}$ (iii) $\cot \Theta$

17. If $i = I_m \sin \left(2\pi ft - \dfrac{\pi}{3}\right)$ determine $\dfrac{d^2i}{dt^2}$.

18. If $v = V_m \cos \left(2\pi ft - \dfrac{\pi}{6}\right)$ determine $\dfrac{d^2V}{dt^2}$.

19. The volume of a sphere is given as $V = \dfrac{4}{3} \pi R^3$, find $\dfrac{d^2V}{dR^2}$.

20. The volume of a cone is given as $V = \dfrac{1}{3} \pi r^2 h$ and $h = 3r$, find $\dfrac{d^2}{dr}$

8.   TANGENTS AND NORMALS

WORKED EXAMPLE 28

Determine the equation of the tangent at a point (1, 2) of the curve $y = 2x^2$.

SOLUTION 28

The equation of a straight line is given by $y = mx + C$ where m is the gradients, and C is the intercept.

The gradient of the curve is found by differentiating
$y = 2x^2$

$$\frac{dy}{dx} = 4x.$$

At $x = 1$, $\dfrac{dy}{dx} = 4$, the equation of the tangent is $y = 4x + c$, this line passes

through the point (1, 2), $x = 1$ and $y = 2$;

$$2 = 4 \times 1 + c$$

$$c = 1 - 2$$

therefore the equation of the line is

$$\boxed{y = 4x - 2}$$

To find the angle between two lines

Consider two lines which intersect at an angle Θ. Let α and β be the angles of the lines that make with the horizontal axis.

$$\tan \alpha = m_1, \text{ and } \tan \beta = m_2$$

$\Theta = \alpha - \beta$, taking tangents on both sides, we have

$$\tan \Theta = \frac{\tan \alpha - \tan \beta}{1 + \tan \alpha \tan \beta} = \frac{m_1 - m_2}{1 + m_1 m_2}.$$

The lines are parallel when $m_1 = m_2 \tan \Theta = 0$.

The lines are perpendicular when $\Theta = 90°$, $\tan 90° = \infty$, that is, $+ m_1 m_2 = 0$

$$\boxed{m_1 m_2 = -1}$$

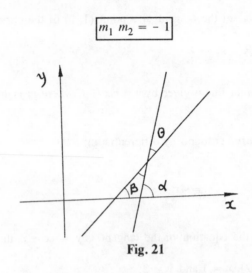

Fig. 21

WORKED EXAMPLE 29

Determine the equation of the normal at the point $(1, 2)$ of the curve $y = 2x^2$.

SOLUTION 29

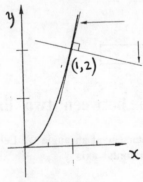

Fig. 22

The equation of the normal is given by the equation $y = mx + c$. If m_1 is the gradient of the tangent m_2 is the gradient of the normal, since the tangent and

normal are at $90°$, $m_1 m_2 = -1$, the gradient of the normal, $m_2 = -\dfrac{1}{m_1} = -\dfrac{1}{4}$.

The equation of the normal is $y = -\dfrac{1}{4}x + c$

this equation passes through the point (1, 2), $\quad 2 = -\dfrac{1}{4} + c, \quad\quad c = 9/4$, then

$y = -\dfrac{1}{4}x + \dfrac{9}{4}\quad 4y + x = 9$ is the equation of the normal.

WORKED EXAMPLE 30

Determine the equations of the tangent and normal to the given curve at the given point adjacent each curve.

(i) $y = 3x^2 + 5$ when $x = -1$
(ii) $y = 3x^2 + 5x - 1$ when $x = 2$
(iii) $y = (x + 1)(x - 3)$ when $x = 1$.

SOLUTION 30

(i) $y = 3x^2 + 5, \quad \dfrac{dy}{dx} = 6x$ at any point,

when $x = -1 \quad \dfrac{dy}{dx} = -6$

The equation of the tangent is given $y = -6x + C$, it passes through the point where $x = -1$ and $y = 3(-1)^2 + 5 = 8, 8 = -6(-1) + C$,
$C = 8 - 6 = 2$

$$\boxed{y = -6x + 2}$$

The gradient of the normal, $m_2 = -\dfrac{1}{m_1} = -\dfrac{1}{-6} = \dfrac{1}{6}$

the equation of the normal $y = \dfrac{1}{6}x + c,$ it passes through $(-1, 8)$, $8 = -\dfrac{1}{6} + c$

$c = 8 + \dfrac{1}{6} = \dfrac{49}{6}$

$$y = \dfrac{1}{6}x + \dfrac{49}{6}$$

the equation of the normal

$$\boxed{6y = x + 49}$$

(ii)　　$y = 3x^2 - 5x - 1$,　$\dfrac{dy}{dx} = 6x - 5$,　the gradient at any point, when $x = 2$,

$\dfrac{dy}{dx} = 6 \times 2 - 5 = 7$.

The equation of the tangent is $y = 7x + c$, it passes through the point $x = 2$

$y = 3(2)^2 - 5(2) - 1 = 12 - 10 - 1 = 1$,

$$2 = 7 \times 1 + c$$

$$c = 2 - 7 = -5$$

$$\boxed{y = 7x - 5}$$

The gradient of the normal is found from $m_1 \, m_2 = -1$, $m_2 = -\dfrac{1}{m_1} = -\dfrac{1}{7}$

$$y = -\frac{1}{7}x + c$$

this passes through the point (2, 1)

$$= -\frac{2}{7} + c$$

$$c = \frac{9}{7}$$

$$y = -\frac{1}{7}x + \frac{9}{7}$$

$$\boxed{7y + x = 9}$$

WORKED EXAMPLE 31

Determine the equations of the tangent and normal to the curve $y = -x^2 - 5x + 6$ at the point when it cuts the x-axis and y-axis.

SOLUTION 31

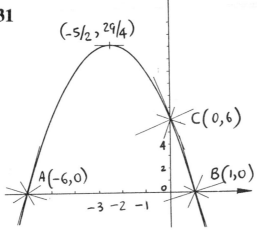

$(-5/2, 29/4)$

$C(0,6)$

$A(-6,0)$

-3 -2 -1

$B(1,0)$

Fig. 23

when $x = 0, y = 1$, the curve has a maximum at $x = -b/2a = -(-5)/2 \ (-1)$

$= -\dfrac{5}{2}, y_{max} = -\left(-\dfrac{5}{2}\right)^2 - 5\left(-\dfrac{5}{2}\right) + 6, \left(-\dfrac{5}{2}, \dfrac{49}{4}\right)$, when $y = 0, x = 1$ and

$x = -6$.

The equations of the tangents and normals at $(-6, 0)$, $(1, 0)$ and $(0, 0)$ are required.

<u>At A (- 6, 0)</u>

The equation of the tangent

$$y = mx + c, \qquad y = \frac{dy}{dx} x + c,$$

$$y = -x^2 - 5x + 6, \qquad \frac{dy}{dx} = -2x - 5$$

when $x = -6$, $\dfrac{dy}{dx} = -2(-6) - 5 = 7$

87

$$y = 7x + c$$

when $x = -6, y = 0, c = 42,$ $\boxed{y = 7x + 42}$

The equation of the normal.
The gradient of the normal is $-1/7$

$$y = -\frac{1}{7}x + c$$

when $x = -6, y = 0, 0 = \dfrac{-(-6)}{7} + c, c = -6/7$

$$y = -\frac{1}{7}x - \frac{6}{7}$$

the equation of the normal

$$\boxed{7y + x + 6 = 0}$$

At B (1, 0)

The equation of the tangent.

$y = -x^2 - 5x + 6,$ $\dfrac{dy}{dx} = -2x - 5,$ when $x = 1, \dfrac{dy}{dx} = -7$

$y = -7x + c,$ when $x = 1, y = 0, c = 7$

$$\boxed{y = -7x + 7}$$

The equation of the normal.

The gradient of the normal is $\dfrac{1}{7}$.

$$y = \frac{1}{7}x + c$$

when $x = 1, y = 0,$ $c = -\dfrac{1}{7}$

$$y = \frac{1}{7}x - \frac{1}{7}$$

$$\boxed{7y = x - 1}$$

At C (0, 6)

The equation of the tangent.

When $x = 0$, $\dfrac{dy}{dx} = -5, y = -5x + c$, when $x = 0, y = 6, c = 6$,

$$\boxed{y = -5x + 6}$$

The equation of the normal

$\dfrac{dy}{dx} = \dfrac{1}{5}, y = \dfrac{1}{5}x + c$, when $x = 0, y = 6, c = 6$

$$\boxed{5y = x + 30}$$

EXERCISES 8

1. The parametric equations of a curve are $x = a(2\Theta - \sin 2\Theta)$ and $y = a(1 - \cos 2\Theta)$.

   Determine $\dfrac{dy}{dx}$ and hence the equation of the tangent and normal to the curve at the point **P** where $\Theta = \pi/4$.

2. Determine the equations of the tangent and normal at any point t on the curve having parametric equations $x = 2 \cos t - \cos 2t, y = 2 \sin t - \sin 2t$.

3. Find the equation of the normal at a general point Θ on the ellipse $x = 2 \cos \Theta$ and $y = 3 \sin \Theta$.

4. Find the equations of the normal and tangent at the point where $x = 5/8$ of the curve whose parametric equations are given by $x = 5 \sin^3 \Theta$, $y = 5 \cos^3 \Theta$.

5. Find the equations of the tangents to the curve $x^2 + y^2 - x - y - 2 = 0$ at the points $(1, -1)$ and $(1, 2)$.

6. Find the equation of the normal at the point $(1, -2)$ of the curve $y^2 = 4x$.

7. Find the equations of the tangent and normal at the point of $(-1, -9)$ of the curve $xy = 9$.

9. SMALL INCREMENTS AND APPROXIMATIONS L'HOPITAL'S RULE RATES OF CHANGE

Small Increments

If $y = f(x)$ and if δx, δy are respectively the increment in x and the corresponding increment in y, the limiting value of the ratio $\delta y/\delta x$ when δx approaches zero is, by definition, dy/dx or $f'(x)$.

$$\frac{\delta y}{\delta x} \rightarrow \frac{dy}{dx} = f'(x) \text{ as } \delta x \rightarrow 0$$

WORKED EXAMPLE 32

The area of a triangle is given as

$$A = = \frac{1}{2} ab \sin \Theta$$

when Θ is the angle between the two sides a, b.

If Θ changes then A changes

$$\frac{dA}{d\Theta} = \frac{1}{2} ab \cos \Theta$$

An error in Θ of $1'$, determine the corresponding error in the area.

SOLUTION 32

$$\frac{\delta A}{\delta \Theta} \approx \frac{dA}{d\Theta} = \frac{1}{2} ab \cos \Theta$$

$$\delta A = \frac{1}{2} ab \cos \Theta \; \delta \Theta$$

If $a = 25$ cm, $b = 45$ cm, $\Theta = 45°$

$$\delta A = \frac{1}{2} \times 25 \times 45 \times \cos 45° \ \delta\Theta$$

$$\delta\Theta = 1^{'} = \left(\frac{1}{60}\right)° = \left(\frac{1}{60} \times \frac{\pi}{180}\right)^0 = 2.9088821 \times 10^{-4}$$

$\delta A = 1/2 \times 25 \times 45 \times \cos 45° \times 2.9088821 \times 10^{-4} = 0.11568$ cm^2.

WORKED EXAMPLE 33

If $y^5 = x^7 (x - 5)$, find using calculus the increase in y when x increases from 3.563 to 3.564, giving your answer correct to four significant figures. State the new value of y in five significant figures.

SOLUTION 33

$y^5 = x^5 (x-5)$, differentiating with respect to x, $5 y^4 \dfrac{dy}{dx} = 5 x^4 (x - 5) + x^5$ which can be written as δy $\delta y = (5 x^5 - 25 x^4 + x^5) \ \delta x$ or $5 y^4 \ \delta y = (6 x^5 - 25 x^4) \ \delta x$

$$\delta y = \frac{6 x^5 - 25 x^4}{5} y^4 \ \delta x = \frac{6 (3.563)^5 - 25 (3.563)^4}{5 \ y^4} (0.001)$$

where $y^5 = (3.563)^7 (3.563 - 5) = -10475.325$

$$y = (-10475.325)^{1/5} = -6.3684465$$

$$\delta y = \frac{\left[6 (3.563)^5 - 25 (3.563)^4\right] \times 0.001}{5 (-6.3684465)^4}$$

$$= \frac{(3445.3263 - 4029.0559) 0.001}{8224.3962}$$

$$= \frac{-0.5837296}{8224.3962} = -7.0975374 \times 10^{-5}$$

92

$= -7.098 \times 10^{-5}$ to four significant figures.

The new value of y is $- 6.3684465 - 7.0975 \times 10^{-5}$ which is equal to 6.3685 to five significant figures.

L' Hôpital's rule

Consider two functions $f(x)$ and $g(x)$ which respresent two curves. If the curves intersect at P when $y = 0$ at $x = a$, then $f(a) = g(a) = 0$.

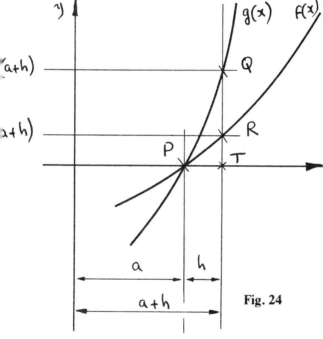

Fig. 24

$x = a + h$, then at Q, $g(x) = g(a + h)$ at R, $f(x) = f(a + h)$

$QT = g(a + h)$, $RT = f(a + h)$

$\dfrac{QT}{RT} = \dfrac{g(a + h)}{f(a + h}$ and dividing by PT, we have

$\dfrac{g(a + h)/h}{f(a + h)/h} = \dfrac{QT/PT}{RT/PT} = \dfrac{\tan QPT}{\tan RPT}$

If $\lim\limits_{x \to a} g(x) / f(x) = \lim\limits_{h \to o} \dfrac{g(a+h)}{f(a+h)} = \lim\limits_{h \to o} = \dfrac{\tan QPT}{\tan RPT} \quad \dfrac{g'(a)}{f'(a)}$

provided $g(x)/(f(x)$ is indeterminate, that is, $g(x)/f(x) = \dfrac{0}{0}$ or $\dfrac{\infty}{\infty}$ and

also provided $g'(a) / f'(a)$ is not indeterminate.

L' Hôpital's rule is only applied when $g(x) / f(x)$ is in the indeterminate form of $\dfrac{0}{0}$ or $\dfrac{\infty}{\infty}$.

WORKED EXAMPLE 34

Find the limit as $\Theta \to 0$ of $\dfrac{\sinh \Theta}{\Theta}$.

SOLUTION 34

$\lim\limits_{\Theta \to 0} \dfrac{\sinh \Theta}{\Theta} = \dfrac{\sinh 0}{0} = \dfrac{0}{0}$

this is indeterminate.

Applying L' Hôpital's rule, that is, differentiating numerator and denominator separately, we have

$\lim\limits_{\Theta \to 0} \dfrac{\sinh \Theta}{\Theta} = \lim\limits_{\Theta \to 0} \dfrac{\cosh \Theta}{1} = \dfrac{\cosh 0}{1}$

$= 1.$

WORKED EXAMPLE 35

If $g(x) = e^x - \cos x$ and $f(x) = 5x$ find the limit as $x \to 0$ of the quotient $g(x) / f(x)$.

SOLUTION 35

$\lim\limits_{x \to 0} \dfrac{e^x - \cos x}{5x} = \dfrac{e^0 - \cos 0}{5\,(0)}$ by direct

substitution, then $\lim\limits_{x \to 0} \dfrac{e^x - \cos x}{5x} = \dfrac{1 - 1}{0} = \dfrac{0}{0}$.

Applying L' Hôpital's rule, that is, differentiating numerator and denominator separately, we have

$\lim\limits_{x \to 0} \dfrac{e^x - \cos x}{5x} = \lim\limits_{x \to 0} \dfrac{e^x + \sin x}{5} \qquad = \dfrac{e^o + \sin(0)}{5} = \dfrac{1}{5}$

therefore $\qquad\qquad \lim\limits_{x \to 0} \dfrac{e^x - \cos x}{5x} = \dfrac{1}{5}$.

WORKED EXAMPLE 36

(i) $\dfrac{\cos 5x - x \sin 3x - 1}{x^3}$

(vii) $\dfrac{\cos (x + \Theta) - \cos (2x - \Theta)}{2x}$

(ii) $\dfrac{4 \cos x + x \sin 2x - 4}{x}$

(viii) $\dfrac{\sin 3x}{3x}$

(iii) $\dfrac{\cos 5x - \sin 7x - 1}{5 \sin 4x}$

(ix) $\dfrac{\cos (x + \pi/2)}{\pi x/2}$

(iv) $\dfrac{\cos x - \cos 3x}{2\,x^2}$

(v) $\dfrac{1 - x + \tan x - \cos x}{x^3}$

(x) $\dfrac{\tan (x - \alpha) + \tan \alpha}{x}$

(vi) $\dfrac{\cos 3x - \cos x}{\sin 2x + \sin x}$

SOLUTION 36

(i) $\lim\limits_{x \to 0} \dfrac{\cos 5x - x \sin 3x - 1}{x^3} = $ indeterminate

applying L, Hôpital's rule, that is, differentiate numeration and denominator separately

$\lim\limits_{x \to 0} \dfrac{-5 \sin 5x - \sin 3x - 3x \cos 3x}{3\,x^2} = \dfrac{0}{0}$

undeterminate, applying the rule again

$\lim\limits_{x \to 0} \dfrac{-25 \cos 5x - 3 \cos 3x - 3 \cos 3x + 3 \times 3x \sin 3x}{6x}$

$= \infty$

(ii) $\lim\limits_{x \to 0} \dfrac{4 \cos x + x \sin 2x - 4}{x} = \dfrac{0}{0}$ indeterminate

applying the rule

$\lim\limits_{x \to 0} \dfrac{-16\,x + \sin 2x + 2\,x \cos 2x}{1} = \dfrac{0}{1} = 0$

(iii) $\lim\limits_{x \to 0} \dfrac{\cos 5x - \sin 7x - 1}{5 \sin 4x} = \dfrac{0}{0}$ indeterminate

$\lim\limits_{x \to 0} \dfrac{-5 \sin 5x - 7 \cos 7x}{20 \cos 4x} = -\dfrac{7}{20}$

(iv) $\lim\limits_{x \to 0} \dfrac{\cos x - \cos 3x}{2\,x^2} = \dfrac{0}{0}$ indeterminate

applying the rule

$$\lim_{x \to 0} \frac{-\cos x + 9 \cos 3x}{4} = 2$$

(v) $\quad \displaystyle\lim_{x \to 0} \frac{1 - x + \tan x - \cos x}{x^3} = \frac{0}{0}$ indeterminate

applying the rule

$$\lim_{x \to 0} \frac{-1 + \sec^2 x + \sin x}{3 x^2} = \frac{0}{0}$$

$$\lim_{x \to 0} \frac{2 \sec^2 x \tan x + \cos x}{6x} = \frac{1}{0} = \infty$$

(vi) $\quad \displaystyle\lim_{x \to 0} \frac{\cos 3x - \cos x}{\sin 2 x + \sin x} = \frac{0}{0}$ indeterminate

$$\lim_{x \to 0} \frac{-3 \sin 3x + \sin x}{2 \cos 2x + \cos x} = \frac{0}{3} = 0$$

(vii) $\quad \displaystyle\lim_{x \to 0} \frac{\cos (x + \Theta) - \cos (2x - \Theta)}{2x} = \frac{0}{0}$ indeterminate

$$\lim_{x \to 0} \frac{-\sin (x + 0) + 2 \sin (2x + 0)}{2} = \frac{\sin \Theta}{2}$$

(viii) $\quad \displaystyle\lim_{x \to 0} \frac{\sin 3x}{3x} = \frac{0}{0}$ indeterminate

$$\lim_{x \to 0} \frac{3 \cos 3x}{3} = 1$$

(ix) $\quad \displaystyle\lim_{x \to 0} \frac{\cos (x + \pi/2)}{\dfrac{\pi}{2} x} = \frac{0}{0}$ indeterminate

$$\lim_{x \to 0} \frac{-\sin (x + \pi/2)}{\pi/2} = -1/(\pi/2) = -2/\pi$$

(x) $\lim\limits_{x \to 0}$ $\dfrac{\tan (x - \alpha) + \tan \alpha}{x} = \dfrac{0}{0}$ indeterminate

$\lim\limits_{x \to 0}$ $\dfrac{\sec^2 (x - \alpha) + \sec^2 \alpha}{1} = 2 \sec^2 \alpha.$

RATE OF CHANGE

Any variable that changes with time is termed as the rate of change.

A distance may change with time, an area may change with time, a volume may change with time, a velocity may change with time.

WORKED EXAMPLE 37

The area of a circle is given by $A = \pi r^2$.

Find the rate of change of A and that of the radius r.

SOLUTION 37

$A = \pi r^2$

differentiating A with respect to r, we have $\dfrac{dA}{dr} = 2\pi r$

dividing dA and dr by dt, we have

$$\dfrac{dA/dt}{dr/dr} = 2\pi r = \dfrac{\text{the rate of change of } A}{\text{the rate of change } r}$$

$$\dfrac{dA}{dr} = 2\pi r \dfrac{dr}{dt}$$

WORKED EXAMPLE 38

The surface area of closed sylindrical can is given by $S = 2\pi rh + 2\pi r^2$. If h is constant find the rates ds/dt and dr/dt.

SOLUTION 38

$S = 2\pi rh + 2\pi r^2$ differentiating with respect to r $\quad \dfrac{dS}{dr} = 2\pi h = 4\pi r$

dividing both D and dr by dt in order to introduce the respective rates

$$\frac{dS/dt}{dr/dt} = 2\pi h + 4\pi r.$$

WORKED SOLUTION 39

The surface and volume of a right circular are given by the expressions

$$S = \pi r (h^2 + r^2)^{1/2} + \pi r^2$$

$$V = \frac{1}{3} \pi r^2 h$$

If $h = 100$ cm, constant, find the rates of S, V when $\quad \dfrac{dr}{dt} = 1$ cm/s

at $r = 10$ cm, to the nearest integer.

$$\frac{dS}{dr} = \pi (h^2 + r^2)^{1/2} + \frac{1}{2} \pi r (h^2 + r^2)^{-1/2} 2r + 2\pi r$$

$$\frac{dS/dt}{dr/dt} = \pi (h^2 + r^2)^{1/2} + \frac{\pi r^2}{(h^2 + r^2)^{1/2}} + 2\pi r \dots (1)$$

$$\frac{dV}{dr} = \frac{2}{3} \pi rh, \frac{dV/dt}{dr/dt} = \frac{2}{2} \pi rh \dots (2)$$

From $\quad (1) \quad \dfrac{dS}{dt} = \left[\pi (100^2 + 10^2)^{1/2} + \dfrac{\pi\, 10^2}{(100^2 + 10^2)^{1/2}} + 2\pi\, 10 \right] 1$

$$\frac{dV}{dr} = \frac{2}{3}\pi rh, \quad \frac{dV/dt}{dr/dt} = \frac{2}{2}\pi rh \quad \dots(2)$$

From (1) $\dfrac{dS}{dt} = \left[\pi \, (100^2 + 10^2)^{1/2} + \dfrac{\pi \, 10^2}{(100^2 + 10^2)^{1/2}} + 2\pi \, 10 \right] 1$

$$= 315.73 + 3.126 + 62.83$$

$$= 382 \text{ cm}^2/\text{s}$$

From (2) $\dfrac{dV}{dt} = \dfrac{2}{3}\pi rh \dfrac{dr}{dt}$

$$= \frac{2}{3}\pi \, (10) \, (100) \, (1)$$

$$= 212 \text{ cm}^3/\text{s}.$$

WORKED EXAMPLE 40

The area of a circle is increasing at the rate of 5 cm^2 per second. Determine the rate of increase of the circumference of the circle when its radius is 6 cm.

$A = \pi r^2$ $\qquad\qquad\qquad\qquad\qquad\qquad C = 2\pi r$

$\dfrac{dA}{dr} = 2\pi r$ $\qquad\qquad\qquad\qquad\qquad \dfrac{dC}{dr} = 2\pi$

$\dfrac{dA/dt}{dr/dt} = 2\pi r$ $\qquad\qquad\qquad\qquad \dfrac{dC/dt}{dr/dt} = 2\pi$

$$2\pi r \, \frac{dC/dt}{dA/dt} = 2\pi$$

$$dC/dt = \frac{dA}{dt} \cdot \frac{1}{r}$$

$\dfrac{dA}{dt} = 5 \text{ cm}^2/\text{s}, \quad \dfrac{dC}{dt} = 5 \times \dfrac{1}{6} = \dfrac{5}{6} \text{ cm/s.}$

WORKED EXAMPLE 41

A uniform soap bubble has a volume $V = \frac{4}{3} \pi r^3$, the rates of the volume to that of the radius is equal to 32π. Determine the radius of the bubble, and the rate of change of the surface area, if the rate of volume increase is 100 cubic units.

SOLUTION 41

$V = \frac{4}{3} \pi r^3$ $\qquad$ $S = \frac{dV}{dr} = 4\pi r^2$

$\frac{dA/dt}{dr/dt} = 32\pi = 4\pi r^2, \frac{dA/dt}{32\pi} dr/dt$ $\qquad r^2 = 8$ $\qquad r = 2\sqrt{2}$ units

$S = 4\pi r^2$ $\qquad \frac{dS}{dr} = 8\pi r$ $\qquad \frac{dS/dt}{dr/dt} = 8\pi r$

$dS/dt = \frac{8\pi r \, dV/dt}{32\pi} = \frac{2\sqrt{2}}{4} dV/dt$

$\qquad = 50\sqrt{2}$.

EXERCISES 9

1. A horizontal trough is 5 m long and has a trapezoidal cross sectional area as shown in **Fig. 25**

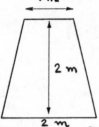

Fig. 25

Water runs into the trough at the rate of 1×10^{-3} m^3 per second. Find the rate at which the water level is rising when the height of the water is 1 m.

2. A spherical balloon has a radius of 10 m. Air is pumped into the balloon at the rate of 5×10^{-3} m^3 per second. Determine the rate at which the radius and the surface area increase.

3. The surface area S and volume V of a solid sphere are changing with respect to time t when it is uniformly heated. When its surface area is increasing at a rate of 1 mm^2/s, determine the rate at which the volume is incrasing when its radious is 10 cm.

4. The volume of an expanding sphere is increasing at the rate 100 mm^3/s. Determine the rate at which the radius of the sphere is increasing at the instant when the radius is 75 mm.

5. The volume of a certain bowl is changing with its height according to the

   equation $\dfrac{dV}{dh} = 3\,e^{3h} + 5\,e^h + 1$.

   If the rate at which the height changes is 3 cm/s determine the rate at which the volume changes when $h = 1$ m.

6. An ellipse has a cartesian equation $4x^2 + 9y^2 = 36$, if x is increasing at the rate of 0.1 cm/sec, find the rate of increase of y.

7. The volume of a cap of a sphere of height h is given by $V = \dfrac{\pi\,h^2}{3}\,(3r - h$

where r is the radius. If $dR/dt = 5$ cm/sec and $h = 20$ cm $=$ constant, find dV/dt when $r = 50$ cm.

8. An equilateral triangle has a side of 10 cm and its perimeter incrases at the rate of 1 cm/s, find the rate of increase of the area. If each side increases to 10.1 cm, find the corresponding increase in the area.

9. Find the limits

(i) $\lim\limits_{x \to 0} \dfrac{\sin 5x}{5x}$

(ii) $\lim\limits_{x \to 0} \dfrac{\tan Kx}{x}$

(iii) $\lim\limits_{x \to 0} \dfrac{2 \sin^{-1} x}{3x}$

(iv) $\lim\limits_{x \to 0} \dfrac{1 - \cos x}{x^2}$

(v) $\lim\limits_{x \to 0} \left(\dfrac{1}{\sin x} - \dfrac{1}{\tan x} \right)$

(vi) $\lim\limits_{x \to \pi/2} \dfrac{\cos x}{\sqrt{(1 - \sin x)^{2/3}}}$

(vii) $\lim\limits_{x \to 0} \dfrac{1 - \cos^3 x}{x \sin 2x}$

(viii) $\lim\limits_{x \to \pi/2} \dfrac{1 - \sin x}{\left(\dfrac{\pi}{2} - x \right)^2}$

(ix) $\lim\limits_{x \to \pi} \dfrac{\sin 3x}{\sin 2x}$

(x) $\lim\limits_{x \to 1} \dfrac{e^x - e}{x - 1}$

(vi) $\lim\limits_{x \to 0} \dfrac{e^{x^2} - \cos x}{x^2}$

(xii) $\lim\limits_{x \to 1} \dfrac{e^x - e^{-x}}{\sin x}$

(xiii) $\lim\limits_{x \to 0} \dfrac{\ln \cos x}{x^2}$

(xiv) $\lim\limits_{x \to 0} \dfrac{1 - \cos (1 - \cos x)}{x^3}$

(xv) $\lim\limits_{x \to 0} \dfrac{1 - \sin (1 + \sin x)}{x^4}$.

10. NEWTON - RAPHSON APPROXIMATION

If x_n is an approximation to a root of $f(x) = 0$ then usually a better approximation is given by x_{n+1}

$$x_{n+1} = x_n - f(x_n) / f'(x_n)$$

Consider a curve that intersects the x-axis at $x = x_n$ as shown in **Fig. 25** the corresponding value of y

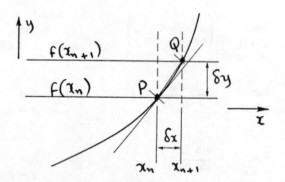

Fig. 26

is $f(x_n)$ which is approximately zero. A better approximation can be achieved that will make the value of y nearer to zero, let this point be at Q which is very close to P, the gradient of the chord PQ is given

$$\frac{\delta y}{\delta x} = \frac{f(x_{n+1}) - f(x_n)}{x_{n+1} - x_n}$$

as δx approaches zero, $\dfrac{\delta y}{\delta x}$ approaches the gradient of the tangent at P, according

to the idea of the limit explained previously. As $\delta x \to 0$, $\dfrac{\delta y}{\delta x} \to \dfrac{dy}{dx} = f'(x_n)$

and $f(x_{n+1}) \to 0$

$$\frac{dy}{dx} = f'(x_n) = -\frac{f(x_n)}{x_{x+1} - x_n}$$

$$x_{n+1} - x_n = -\frac{f(x_n)}{f'(x_n)}$$

$$\boxed{x_{n+1} = x_n - \frac{f(x_n)}{f'(x_n)}}$$

WORKED EXAMPLE 42

Plot the graph of the cubic function $y = 2x^3 - 3x^2 - 11x - 12$ between the values $x = -4$ and $x = 4$, at intervals of 0.5. Give your answers correct to two decimal places, for the equation $2x^3 - 3x^2 - 11x + 8 = 0$.

SOLUTION 42

x	- 4	- 3.5	- 3.0	- 2.5
	0.5	1	1.5	2
- 12	- 12	- 12	- 12	- 12
	- 12	- 12	- 12	- 12
- 11x	44	38.5	33	27.5
	- 5.50	- 11	- 16.5	- 22
- 3 x^2	- 48	- 36.75	- 27	- 18.75
	- 0.75	- 3	- 6.75	- 12
2 x^3	- 128	- 85.75	- 54	- 31.25
	0.25	2	6.75	16
y	- 144	- 96	- 60	- 34.5
	- 18.00	- 24	- 28.5	- 30

contd ...

x	- 2.0	- 1.5	- 1.0	- 0.5	0
	2.5	3	3.5	4	
- 12	- 12	- 12	- 12	- 12	- 12
	- 12	- 12	- 12	- 12	
- 11x	22	16.5	11	5.5	0
	- 27.5	- 33	- 38.5	- 44	
$- 3x^2$	- 12	- 6.75	- 3	- 0.75	0
	- 18.75	- 27	- 36.25	- 48	
$2x^3$	- 16	- 675	- 2	- 0.25	0
	31.25	54	85.75	128	
y	- 18	- 9	- 6	- 7.5	- 12
	- 27	- 18	- 1.524	24	

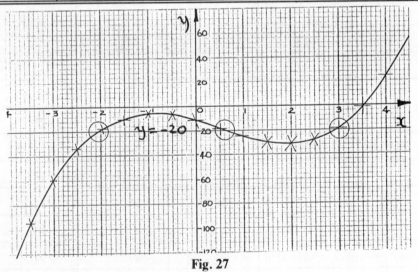

Fig. 27

106

Use the following scales on the x-axis 1 cm = 0.5 and the y-axis 1 cm = 10.

y is plotted against x and a smooth curve is obtained with a maximum point at $x = -1$ and a minimum point at $x = 2$, the graph cuts the x-axis at $x = 3.55$ and the y-axis at $y = -12$ in order to solve the equation $2x^3 - 3x^2 - 11x + 8 = 0$, we make $y = -20$, a horizontal line intersects the graph at $x = -2.08, x = 0.68$ and $x = 2.93$.

$$y = 2x^3 - 3x^2 - 11x - 12 = -20.$$

Let $f(x) = 2x^3 - 3x^2 - 11x + 8 \ldots(1)$.

We would like to find better approximations for the roots
$x_{n_1} = -2.08$, $x_{n_2} = 0.68$ and $x_{n_3} = 2.93$.

Let us apply the Newton-Raphson method of approximation.

$$x_{n_1 + 1} = x_{n_1} - \frac{f\left(x_{n_1}\right)}{f'\left(x_{n_1}\right)}$$

$$f\left(x_{n_1}\right) = 2(-2.08)^3 - 3(-2.08)^2 - 11(-2.08) + 8$$

$$= -17.998 - 12.98 + 27.88 + 8$$
$$= -0.098.$$

Differentiating equation (1), we have $f'(x) = 6x^2 - 6x - 11$ for $x = -2.08$

$$f'\left(x_{n_1}\right) = 6(-2.08)^2 - 6(-2.08) - 11$$

$$= 25.9584 + 12.48 - 11$$
$$= 27.4384.$$

A better approximation is $x_{n_1 + 1}$

$$x_{n_1 + 1} = x_{n_1} - \frac{f(x_{n_1})}{f'(x_{n_1})} = -2.08 - \frac{(-0.098)}{27.4384}$$

$$= -2.08 + 0.0035716$$

$$= -2.076.$$

This may be further improved if the method is repeated.

$$f x_{n_1 + 1} = 2(-2.076)^3 - 3(-2.076)^2 - 11(-2.076) + 8$$

$$= -17.89419 - 12.929328 + 22.836 + 8$$
$$= 0.012482$$
$$= 0.0125 \text{ to three decimal places}$$
$$f'(-2.076) = 6(-2.076)^2 - 6(-2.076) - 11$$
$$= 25.958656 + 12.456 - 11$$
$$= 27.314656$$

$$(x_{n_1 + 2}) = (x_{n_1 + 1}) - \frac{f(x_{n_1 + 1})}{f'(x_{n_1 + 1})} = -2.076 - \frac{0.0125}{27.314656}$$

$$= -2.076 - 0.00045762978$$

$$x_{n_1 + 2} = -2.0765$$

$$x_{n_1 + 2} = -2.077 \text{ to four significant figures}$$
$$= -2.08 \text{ to three significant figures}$$
$$= -2.1 \text{ to two significant figures.}$$

Let us apply again the Newton-Raphson method of approximation for the second root, $x_{n_2} = 0.68$

$$f(x) = 2x^3 - 3x^2 - 11x + 8$$
$$f'(x) = 6x^2 - 6x - 11$$
$$f(x_{n_1}) = 2(0.68)^3 - 3(0.68)^2 - 11(0.68) + 8$$
$$= 0.628864 - 1.3872 - 7.48 + 8$$
$$= -0.238$$
$$f'(x_{n_1 + 1}) = 6(0.68)^2 - 6(0.68) - 11$$

$$= -12.3056$$

$$x_{n_2 + 1} = x_{n_2} - \frac{f(x_{n_2})}{f'(x_{n_2})} = 0.68 - \frac{(-0.238)}{-12.3056}$$

$$x_{n_2 + 1} = 0.661$$

$$= 0.66 \text{ to two decimal places.}$$

Let us apply finally the Newton-Raphson method to the third root, $x_{n_3} = 2.93$.

$$f(x) = 2x^3 - 3x^2 - 11x + 8$$
$$f'(x) = 6x^2 = 6x - 11, f'(x_{n_3}) = 6(2.93)^2 - 6(2.93 - 11 = 22.9294$$

$$f(x_{n_3}) = 2 (2.93)^3 - 3 (2.93)^2 - 11(2.93) + 8$$

$$= 50.307514 - 25.7547 - 32.23 + 8$$
$$= 0.322814.$$

A better approximation is as follows:-

$$x_{n_3 + 1} = x_{n_3} - \frac{f(x_{n_3})}{f'(x_{n_3})} = 2.93 - \frac{0.322814}{22.9294}$$

$$= 2.9159214$$

$= 2.92$ to three significant figures.

The roots of the cubic equation $2 x^3 - 3 x^2 - 11x + 8 = 0$ are now improved to the values $- 2.08, 0.66$ and 2.92 to three significant figures.

WORKED EXAMPLE 43

Taking $x = 43.5°$ as a first approximation to a root of the equatoon $\sin x + \sin \frac{x}{2} - = 0$

apply the Newton-Raphson procedure to find a further approximation giving your answer to 3 significant figures.

SOLUTION 43

$$f(x) = \sin x + \sin \frac{1}{2} x - 1$$

$$f'(x) = \cos + \frac{1}{2} \cos \frac{1}{2} x$$

$x_n = 43.5°$, it is required to find $x_{n + 1}$, a better approximation.

$$f(x_n) = \sin 43.5° + \sin \frac{1}{2} (43.5°) - 1$$

$$= 0.6883545 + 0.3705574 - 1$$
$$= 0.0591024$$

$$f'(x_n) = \cos 43.5° + \frac{1}{2} \cos \frac{1}{2} (43.5°)$$

$$= 1.1897791$$

$$x_{n+1} = x_n - \frac{f(x_n)}{f'(x_n)} = 43.5 \times \frac{\pi}{180} - \frac{0.0591024}{1.1897791} = 0.759218 - 0.0496751$$

$$= 0.7095431° \times \frac{180}{\pi} = 40.653826°$$

$$= 40.7°$$

WORKED EXAMPLE 44

Taking $x = 0.324^c$ as a first approxmation to a root of the equation
$2 \cos (2x - 2\pi/3) - x = 0$ apply the Newton-Raphson procedure to find two further
approximations giving your answers to 3 decimal places.

SOLUTION 44

$f(x) = 2 \cos (2x - 2\pi/3) - x$
$f'(x) = -4 \sin (2x - 2\pi/3) - 1$

When $x_n = 0.324^c$ (note the angle is in radians)
$f(x_n) = 2 \cos (2 \times 0.324 - 2\pi/3)^c - 0.324^c$
$\qquad = -0.0758387$
$f'(x_n) = -4 \sin (2 \times 0.0324 - 2\pi/3)^c - 1$
$\qquad = 2.9690886.$

A better approximation is found as follows:-

$$x_{n+1} = x_n - \frac{f(x_n)}{f'(x_n)} = 0.324 - \frac{(-0.0758387)}{2.9690886}$$

$$= 0.3495427 = 0.350^c \text{ to three decimal places.}$$

A better still approximation is found as follows:-

$$x_{n+2} = x_{n+1} - \frac{f(x_{n+1})}{f'(x_{n+1})}$$

$$f(x_{n+1}) = 2 \cos (2 \times 0.350 - 2\pi/3) - 0.350$$
$$= 0.00097557814$$
$$f'(x_{n+1}) = -4 \sin (2 \times 0.350 - 2\pi/3) - 1$$
$$= 2.9379264$$

$$x_{n+2} = x_{n+1} - \frac{f(x_{n+1})}{f'(x_{n+1})} = 0.350 - \frac{0.00097557814}{2.9379264}$$

$$= 0.3496679 = 0.350^c \text{ to three decimal places.}$$

The two answers are correct to three decimal places.

WORKED EXAMPLE 45

The graphs of $y = \tan^{-1} x$ and $x/5 + y/6 = 1$ are drawn giving points of intrsections to the first approximations of $(3.8, 1.313^c)$ and $(1.55, 4.14^c)$.

Find better approximations to these solutions by applying Newton-Raphson formula.

SOLUTION 45

$$y = \tan^{-1} x, \quad \frac{x}{5} + \frac{y}{6} = 1$$
$$y = \tan^{-1} x = 6\left(1 - \frac{x}{5}\right) = 6 - \frac{6}{5}x$$

$$5 \tan^{-1} x + 6x - 30 = 0$$

$$f(x) = 5 \tan^{-1} x + 6x - 30$$

$$f'(x) = \frac{5}{1 + x^2} + 6$$

$$f\left(x_{n_1}\right) = 5 \tan^{-1} x_{n_1} + 6 x_{n_1} - 30$$

and $f'\left(x_{n_1}\right) = \dfrac{5}{1 + x_{n_1}^{2}} + 6.$

for $(3.8, 1.313^c)$, that is, $x_{n_1} = 3.8$ and $\tan^{-1} x_{n_1} = 1.313$.
$f\left(x_{n_1}\right) = 5 \times 1.313 + 6 \times 3.8 - 30 = 29.365 - 30 = -0.635$

$$f'\left(x_{n_1}\right) = \dfrac{5}{1 + x_{n_1}^{2}} + 6 = \dfrac{5}{1 + 3.8^2} + 6 = 6.3238342$$

$$x_{n_1 + 1} = x_{n_1} - \dfrac{f\left(x_{n_1}\right)}{f'\left(x_{n_1}\right)} = 3.8 \quad \dfrac{(-0.635)}{6.3238342} = 3.90.$$

For $(1.55, 4.14^c)$. that is, $x_{n_2} = 1.55$ and $\tan^{-1} x_{n_2} = 4.14^c$

$$f\left(x_{n_2}\right) = 5 \tan^{-1} x_{n_2} + 6 x_{n_2} - 30 \; 5 \times 4.14 + 6 \times 1.55 - 30 = 0$$

$$f'\left(x_{n_2}\right) = \dfrac{5}{1 + x_{n_2}^{2}} + 6 = \dfrac{5}{1 + 1.55^2} + 6 = 7.4695077$$

$$x_{n_2 + 1} = x_{n_2} - \dfrac{f\left(x_{n_2}\right)}{f'\left(x_{n_2}\right)} = 1.55 - \dfrac{0}{7.4695077} = 1.55$$

Therefore $x_{n_2 + 1} = 1.55.$

The two better approximations are:-

$$x_{n_1 + 1} = 3.90 \quad \text{and} \quad x_{n_2 + 1} = 1.55.$$

EXERCISES 10

1. Show graphically, or otherwise that the cubic equations

   (i) $x^3 + 3x + 3 = 0$ (ii) $x^3 + 3x + 28 = 0$

   have only one real root, and prove that this root lies between -0.7 and -0.9 for (i) and between -2.6 and -2.8 for (ii).

   By taking $x = -0.75$ for (i) and $x = -2.7$ for (ii) as first approximations to these roots, giving your answer to 3 significant figures.

2. Find graphically the solution of the equation $f(x) = 2x - e^{-x} = 0$. (The graphs of $y = 2x$ and $y = e^{-x}$ are plotted, the intersection gives the solution $x = 0.35$).

   Taking $x = 0.35$ as a first approximation and using the Newton-Raphson procedure to the root of the equation $f(x) = 0$ giving your answers to 4 decimal places.

   Solve the equations $3x - e^x = 0$, $4x - e^x = 0$ graphically and using the Newton-Raphson method to obtain improved solutions.

3. Taking $x = 1.414^c$ as a first approximation to a root of the equation

   $(\cos 2x - 2\pi/3) = \dfrac{1}{2} x$ apply the Newton-Raphson procedure to find two

   further approximations giving your answers to three decimal places.

11. MACLAURIN'S EXPANSIONS

Power Series

The following are some examples of infinite series, a series in which the number of terms is allowed to increase without a limit.

1.
$$(1 + x)^n = 1 + nx + \frac{n(n-1)}{2!}x^2 + \frac{n(n-1)(n-2)}{3!}x^3$$

$$+ \ldots + \frac{n(n-1)\ldots(n-r+}{}$$

(n is not a positive integer, $|x| < 1$).

2. $\quad e^x = \exp x = 1 + x + \frac{x^2}{2!} + \frac{x^3}{3!} + \ldots + \frac{x^r}{r!} + \ldots$ for all values of x

3. $\quad \ln(1 + x) = x - \frac{x^2}{2} + \frac{x^3}{3} - \ldots + (-1)^{r+1}\frac{x^r}{r} + \ldots (-1 < x \le 1)$

4. $\quad \cos x = 1 - \frac{x^2}{2!} + \frac{x^4}{4!} - \ldots + (-1)^r\frac{x^{2r}}{2r)!} + \ldots$ for all values of x

5. $\quad \sin x = x - \frac{x^3}{3!} + \frac{x^5}{5!} - \ldots + (-1)^r\frac{x^{2r+1}}{(2r+1!)} + \ldots$ for all values of x

6. $\quad \cosh x = 1 + \frac{x^2}{2!} + \frac{x^4}{4!} + \ldots + \frac{x^{2r}}{(2r)!} + \ldots$ for all values of x

7. $\quad \sinh x = x + \frac{x^3}{3!} + \frac{x^5}{5!} + \ldots + \frac{x^{2r+1}}{(2r+1)!} + \ldots$ for all values of x.

WORKED EXAMPLE 46

Use the expansions of e^x and e^{-x} to obtain the following expansions:-

(i) cosh x (ii) sin x (iii) cos x (iv) sin x.

$$\frac{1)}{r!}x^r + \ldots$$

SOLUTION 46

(i) $\cosh x = \dfrac{1}{2}\left[e^x + e^{-x}\right] = \dfrac{1}{2}\left[1 + x + \dfrac{x^2}{2!} + \ldots + 1 - x + \dfrac{x^2}{2!} - \ldots\right]$

$= \dfrac{1}{2}\left[2 + 2\dfrac{x^2}{2!} + 2\dfrac{x^4}{4!} + \ldots\right] = 1 + \dfrac{x^2}{2!} + \dfrac{x^4}{4!} + \ldots$

(ii) $\sinh x = \dfrac{1}{2}\left[e^x - e^{-x}\right] = \dfrac{1}{2}\left[1 + x + \dfrac{x^2}{2!} + \ldots - \left(1 - x + \dfrac{x^2}{2!} - \ldots\right)\right]$

$= \dfrac{1}{2}\left[2x + 2\dfrac{x^3}{3!} + \ldots\right] = x + \dfrac{x^3}{3!} + \ldots$

(iii) $\cos x = \dfrac{1}{2}\left[e^{ix} + e^{-ix}\right] = \dfrac{1}{2}\left[\left(1 + ix + \dfrac{i^2 x^2}{2!} + \ldots\right) + \left(1 - ix + \dfrac{i^2 x^2}{2!} - \ldots\right)\right]$

$= 1 - \dfrac{x^2}{2!} + \dfrac{x^4}{4!} - \ldots$

(iv) $\sin x = \dfrac{e^{ix} - e^{-ix}}{2} = \dfrac{1}{2!}\left[\left(1 + ix + \dfrac{i^2 x^2}{2!} + \ldots\right) - \left(1 - ix + \dfrac{i^2 x^2}{2!} - \ldots\right)\right]$

$= x - \dfrac{x^3}{3!} + \dfrac{x^5}{5!} - \ldots$

WORKED EXAMPLE 47

Use the expansions of $\sin x$ and $\cos x$ to obtain the following expansions:-

(i) $\tan x$ and (ii) $\tanh x$, as far as the term in x^5.

SOLUTION 47

(i) $\tan x = \dfrac{\sin x}{\cos x} = \dfrac{x - x^3/3! + x^5/5! - \ldots}{1 - x^2/2! + x^4/4! - \ldots}$

Using long division, we have

$$x + \dfrac{1}{3}x^3 + \dfrac{2}{15}x^5 + \ldots$$

$1 - x^2/2! + x^4/4! - \ldots$ $\bigg|$ $x - x^3/3! + x^5/5! - \ldots$
$x - x^3/2! + x^5/4! - \ldots$

$$\left(-\frac{1}{6} + \frac{1}{2}\right)x^3 + \left(\frac{1}{120} - \frac{1}{24}\right)x^5$$

$$= \frac{1}{3}x^3 - \frac{1}{30}x^5$$

$$\frac{1}{3}x^3 - \frac{1}{6}x^5$$

$$\left(-\frac{1}{30} + \frac{1}{6}\right)x^5$$

$$= \frac{2}{15}x^5$$

$$\tan x = x + \frac{1}{3}x^3 + \frac{2}{15}x^5$$

(ii) $\qquad \tanh x = \dfrac{\sinh x}{\cosh x} = \dfrac{x + x^3/3! + x^5/5! + \ldots}{1 + x^2/2! + x^4/4! + \ldots}$

using long division, we have

$$x - \frac{1}{3}x^3 + \frac{2}{14}x^5$$

$1 + x^2/2! + x^4/4! + \ldots$ $\bigg|$ $x + x^3/3! + x^5/5! + \ldots$
$x + x^3/2! + x^5/4! + \ldots$

$$\left(\frac{1}{6} - \frac{1}{2}\right)x^3 + \left(\frac{1}{120} - \frac{1}{24}\right)x^5$$

$$= -\frac{1}{3}x^3 - \frac{1}{30}x^5$$

$$-\frac{1}{3}x^3 - \frac{1}{6}x^5$$

116

$$\left(-\frac{1}{30} + \frac{1}{6}\right) x^5$$

$$= \frac{2}{15} x^5$$

$$\tanh x = x - \frac{1}{3} x^3 + \frac{2}{15} x^5.$$

WORKED EXAMPLE 48

Expand $e^{\sin x}$ as far as the term in x^4.

SOLUTION 48

$$e^{\sin x} = e^{x - x^3/3! + x^5/5! - \ldots}$$

$$= e^x \, e^{-x^3/3!} \, e^{x^5/5!} \, e^{-x^7/7!} \ldots$$

$$= \left(1 + x + \frac{x^2}{2!} + \frac{x^3}{3!} + \frac{x^4}{4!}\right) \left(1 - \frac{x^3}{3!}\right) \left(1 + \frac{x^5}{5!}\right) \left(1 - \frac{x^7}{7!}\right)$$

$$= 1 + x + \frac{x^2}{2!} + \frac{x^3}{3!} + \frac{x^4}{4!} - \frac{x^3}{3!} - \frac{x^4}{3!}$$

$$= 1 + x + \frac{1}{2} x^2 \left(\frac{1}{24} - \frac{1}{6}\right) x^4$$

$$= 1 + x + \frac{1}{x} x^2 - \frac{1}{8} x^4$$

Taylor's theorem states:-

$$f(x + h) = f(x) + h f'(x) + \frac{h^2}{2!} f''(x) + \frac{h^3}{3!} f'''(x) + \ldots$$

Let $f(x + h) = a_0 + a_1 h + a_2 h^2 + a_3 h^3 + \ldots$ \hfill (1) where $a_0, a_1, a_2, a_3 \ldots$ are functions of x.

Differentiating with respect to h, where $u = x + h$

$$\frac{d}{dh} f(x + h) = \frac{d}{du} f(u) \frac{du}{dh}, \frac{du}{dh} = 1$$

$$= f^{I}(u)$$

$$f^{I}(u) = a_1 + 2 a_2 h + 3 a_3 h^2 + 4 a_4 h^3 \qquad (2)$$
$$f^{II}(u) = 2 a_2 + 3 \times 2 a_3 h + 4 \times 3 a_4 h^2 + \ldots \qquad (3)$$
$$f^{III}(u) = 3 \times 2 \times 1 a_3 + 4 \times 3 \times 2 a_4 h + \ldots \qquad (4)$$

If $h = 0$, from (1) we have $f \quad (x) = a_0$
from (2) we have $f^{I} \quad (x) = a_1$
from (3) we have $f^{II} \quad (x) = a_2$
from (4) we have $f^{III} \quad (x) = a_3$

Therefore,

$$f(x + h) = f(x) + h f^{I}(x) + \frac{h^2}{2!} f^{II}(x) \frac{h^3}{3!} f^{III}(x) + \ldots$$

WORKED EXAMPLE 49

Prove the binomial expansion

$$(a + b)^n = a^n + n a^{n-1} b + \frac{n(n-1)}{2!} a^{n-2} b^2 + \frac{n(n-1)(n-2)}{3!} a^{n-3} b^3 +$$

using Taylor's theorem.

SOLUTION 49

Let $f \quad (a) = a^n$
$f^{I} \quad (a) = n a^{n-1}$
$f^{II} \quad (a) = n(n-1) a^{n-2}$
$f^{III} (a) = n(n-1)(n-2) a^{n-3}$

....

Using Taylor's theorem which states

$$f(x + h) = f(x) + h f'(x) + \frac{h^2}{2!} f''(x) + \frac{h^3}{3!} f'''(x) + \ldots$$

If $= a$ and $h = B$

$$(a + b)^n = f(a) + b f'(a) + \frac{b^2}{2!} f''(a) + \frac{b^3}{3!} f'''(a) + \ldots$$

$$(a + b)^n = a^n + n a^{n-1} b + \frac{n(n-1)}{2!} a^{n-2} b^2 + \frac{n(n-1)(n-2)}{3!} a^{n-3} b^3 + \ldots$$

Successive approximations

Consider the function $f(x) = \dfrac{1}{1 - x}$

where x is less than unity.

Using long division, we have

$$
\begin{array}{r}
1 + x + x^2 \\ \hline
1 - x \quad | \quad 1 \\
1 - x \\ \hline
x \\
x - x^2 \\ \hline
x^2 \\
x^2 - x^3 \\ \hline
x^3
\end{array}
$$

It is observed that

$$\frac{1}{1 - x} = 1 + \frac{x}{1 - x}$$

$$\frac{1}{1 - x} = 1 + x + \frac{x^2}{1 - x}$$

$$\frac{1}{1-x} = 1 + x + x^2 + \frac{x^3}{1-x}$$

Hence $1, 1 + x, 1 + x + x^2, \dots$ are successive aproximations to the function $f(x)$ and the respective errors are $\dfrac{x}{1-x}, \dfrac{x^2}{1-x}, \dfrac{x^3}{1-x}, \dots,$

If $x = 0$, the successive approximations are all equal.

Therefore, the function can be written as $f(x) = a_0 + a_1 x + a_2 x^2 + a_3 x^3 + \dots.$

Maclaurin's theorem states:

$$f(x) = f(0) + x f'(0) + \frac{x^2}{2!} f''(0) + \dots$$

Let $(lf)(x) = a_0 + a_1 x + a_2 x^2 + a_3 x^3 + a_4 x^4 + \dots$
where $a_0, a_1, a_2, a_3, \dots$ are constants.

Differentiating with respect to x (1)

$$f'(x) = a_1 + 2 a_2 + 3 a_3 x^2 + 4 a_4 x^3 + \dots (2)$$

Differentiating with respect to x (2)

$$f''(x) = 2 + 3 \times 2 a_3 x + 4 \times 3 a_4 x^2 + \dots (3)$$

Differentiating with respect to x (3)

$$f'''(x) = 3 \times 2 \times 1 a_3 + 4 \times 3 \times 2 a_4 x + (4)$$

If $x = 0$, (1), (2), (3), (4), $\dots$ become

$$a_0 = f(0)$$
$$a_1 = f'(0)$$
$$a_2 = f''(0)$$
$$a_3 = f'''(0)$$

Therefore,

$$f(x) = f(0) + x f'(0) + \frac{x^2}{2!} f''(0) + \frac{x^3}{3!} f'''(0) + + \frac{x^r}{r!} f^r(0) + \dots$$

Observe that Maclaurin's theorem is the special case of Taylor's theorem where x is replaced by zero and h by x.

WORKED EXAMPLE 50

Express the following functions:-

(i) cosh x (ii) sinh x (iii) cos x (iv) sin x (v) ln $(1 + x)$ as power series using Maclaurin's theorem as far as the term in x^4.

SOLUTION 50

(i)

$$f(x) = \cosh x \qquad f(1) = 1$$
$$f'(x) = \sin x \qquad f'(0) = 0$$
$$f''(x) = \cosh x \qquad f''(0) = 1$$
$$f'''(x) = \sin x \qquad f'''(0) = 0$$
$$f^{iv}(x) = \cosh x \qquad f^{iv}(0) = 1$$

$$\cosh x = 1 + \frac{x^2}{2!} + \frac{x^4}{4!}.$$

(ii)

$$f(x) = \sinh x \qquad f(0) = 1$$
$$f'(x) = \cosh x \qquad f'(0) = 1$$
$$f''(x) = \sinh x \qquad f''(0) = 0$$
$$f'''(x) = \cosh x \qquad f'''(0) = 1$$
$$f^{iv}(x) = \sinh x \qquad f^{iv}(0) = 0$$

$$\sinh x = x + \frac{x^3}{3!}.$$

(iii)

$$f(x) = \cos x \qquad f(0) = 1$$
$$f'(x) = \sin x \qquad f'(0) = 0$$
$$f''(x) = \cos x \qquad f''(0) = 1$$
$$f'''(x) = \sin x \qquad f'''(0) = 0$$
$$f^{iv}(x) = \cos x \qquad f^{iv}(0) = 1$$

$$\cos x = 1 - \frac{x^2}{2!} + \frac{x^4}{4!}.$$

(iv)
$$f \ (x) = \sin x \qquad\qquad f \ (0) = \quad 0$$
$$f' \ (x) = \cos x \qquad\qquad f' \ (0) = \quad 1$$
$$f'' \ (x) = \sin x \qquad\qquad f'' \ (0) = \quad 0$$
$$f''' \ (x) = \cos x \qquad\qquad f''' \ (0) = -1$$
$$f^{\ iv} \ (x) = \sin x \qquad\qquad f^{\ iv} \ (0) = \quad 0$$

$$\sin x = x - \frac{x^3}{3!}.$$

(v)
$$f \quad (x) = \ln (1 + x \qquad f \ (0) = \quad 0$$

$$f' \ (x) = \quad \cos \quad \frac{1}{1 + x} \qquad f \ (0) = \quad 1$$

$$f'' \ (x) = \quad \frac{1}{(1 + x)^2} \qquad f'' \ (0) = -1$$

$$f''' \ (x) = \quad \frac{2}{(1 + x)^3} \qquad f''' \ (0) = \quad 2$$

$$f^{\ iv} \ (x) = \quad \frac{-6}{1 + x)^4} \qquad f^{\ iv} \ (0) = -6$$

$$\ln (1 + x) = x - \frac{x^2}{2} + \frac{x^3}{3} - \frac{x^4}{4}.$$

WORKED EXAMPLE 51

Find the first five terms in the expansion of ln $(1 + \sin x)$, using Maclaurin's theorem.

SOLUTION 51

$$f (x) \ = \ln (1 + \sin x) \qquad\qquad f \ (0) = 0$$

$$f' (x) = \quad \frac{\cos x}{1 + \sin x} \qquad\qquad f' (0) = 1$$

$$f'' (x) = \quad \frac{- \sin x) (1 + \sin x) - \cos^2 x}{1 + \sin x)^2}$$

$$= \quad \frac{- \sin x - \sin^2 x - \cos^2 x}{(1 + \sin x)^2}$$

$$= \quad \frac{1 + \sin x}{(1 + \sin x)^2} = - \frac{1}{1 + \sin x} \qquad f'' (0) = -1$$

122

$$f'''(x) = \frac{\cos x}{(1 + \sin x)^2} \qquad f'''(0) = 1$$

$$f^{iv}(x) = \frac{- \sin x (1 + \sin x)^2 - 2 \cos^2 x \ (1 + \sin x)}{(1 + \sin x)^3}$$

$$f^{v}(x) = \frac{- \sin x (1 + \sin x) - 2 \cos^2 x}{(1 + \sin x)^3}$$

$$f^{vi}(x) = \frac{- \sin x - \sin^2 x - 2 \cos^2 x}{(1 + \sin x)^3} \qquad f^{iv}(0) = -2$$

$$f^{v}(x) = (- \cos x - 2 \sin x \cos x \_ 4 \cos x \sin x)(1 + \sin x)^3$$

$$\frac{+ (\sin x + \sin^2 x + 2 \cos^2 x) \, 3 \, (1 + \sin x)^2 \cos x}{(1 + \sin x)^6}$$

$$f^{v}(0) = \frac{-1 + 6}{1} = 5$$

$$\ln (1 + \sin x) = x - \frac{x^2}{2} + \frac{x^3}{6} = \frac{x^4}{12} + \frac{x^5}{24}.$$

WORKED EXAMPLE 52

If $\sin x = x - \dfrac{x^3}{3!} + \dfrac{x^5}{5!} - \dfrac{x^7}{7!} + \ldots$

find the expansion for $\cos x$.

SOLUTION 52

$$\sin x = x - \frac{x^3}{3!} + \frac{x^5}{5!} - \frac{x^7}{7!} + \ldots$$

Differentiating with respect to x.

$$\frac{d}{dx} (\sin x) = 1 - \frac{3 x^2}{3!} + \frac{5 x^4}{5!} - \frac{7 x^6}{7!} + \ldots$$

$$\cos x = 1 - \frac{x^2}{2!} + \frac{x^4}{4!} - \frac{x^6}{6!} + \ldots$$

WORKED EXAMPLE 53

Find the expansions of the following functions:-

(i) $\dfrac{1}{x^2 - 9}$ (ii) $\dfrac{1}{4 - x^2}$ (iii) $\dfrac{1}{\sqrt{5^2 - x^2}}$

using Maclaurin's theorem as far as the term in x^3.

SOLUTION 53

(i) $f(x) = \dfrac{1}{x^2 - 9}$ $f(0) = -1/9$

$$f'(x) = - \frac{2x}{(x^2 - 9)^2} \qquad f'(0) = 0$$

$$f''(x) = - \frac{2(x^2 - 9)^2 - 2x\,2\,(x^2 - 9)\,2x}{(x^2 - 9)^4}$$

$$= - \frac{2(x^2 - 9) - 8x^2}{(x^2 - 9)^3} = \frac{+18 + 6x^2}{(x^2 - 9)^3}$$

$$f''(0) = \frac{18}{-9^3}$$

$$f''(0) = - \frac{18}{93} = - \frac{2}{81}$$

$$f'''(x) = \frac{12x\,(x^2 - 9)^3 - (18 + 6x^2)\,3\,(x^2 - 9)^2\,(2x)}{(x^2 - 9)^6}$$

$$= \frac{12x\,(x^2 - 9) - 6x\,(18 + 6x^2)}{(x^2 - 9)^4} \qquad f'''(0) = 0$$

$$\frac{1}{x^2 - 9} = - \frac{1}{9} - \frac{2}{81}\frac{x^2}{2} = - \frac{1}{9} - \frac{x^2}{81}.$$

(ii) $f(x) = \dfrac{1}{4 - x^2}$ $f(0) = \dfrac{1}{4}$

$$f'(x) = -\frac{1 \times (-2x)}{(4 - x^2)^2} = \frac{2x}{(4 - x^2)^2}, f'(0) = 0$$

$$f''(x) = \frac{2(4 - x^2)^2 - 2 \times 2(4 - x^2)(-2x)}{(4 - x^2)^4}$$

$$= \frac{2(4 - x^2) + 8x^2}{(4 - x^2)^3} = \frac{8 - 2x^2 + 8x^2}{(4 - x^2)^3}$$

$$= \frac{8 + 6x^2}{(4 - x^2)^3}, f''(0) = \frac{8}{4^3} = \frac{1}{8}$$

$$f'''(x) = \frac{12x(4 - x^2)^3 - (8 + 6x^2)\,3(4 - x^2)^2(-2x)}{(4 - x^2)^4}$$

$$= \frac{12x(4 - x^2) + 6x(8 + 6x^2)}{(4 - x^2)^4}$$

$$\boxed{f'''(0) = 0}$$

$$\frac{1}{4 - x^2} = \frac{1}{4} + \frac{1}{8}\frac{x^2}{2} = \frac{1}{4} + \frac{x^2}{16}.$$

(iii) $\dfrac{1}{\sqrt{5^2 - x^2}} = (25 - x^2)^{-1/2} = f(x)$ $f(0) = \dfrac{1}{5}$

$$f'(x) = -\frac{1}{2}(25 - x^2)^{-3/2}(-2x) = x(25 - x^2)^{-3/2} \qquad f'(0) = 0$$

$$f''(x) = (25 - x^2)^{-3/2} - \frac{3}{2}x(25 - x^2)^{-5/2}(-2x) \qquad f''(0) = \frac{1}{125}$$

$$f''(x) = (25 - x^2)^{-3/2} + 3x^2(25 - x^2)^{-5/2}$$

$$f'''(x) = -\frac{3}{2}(25 - x^2)^{-5/2}(-2x) + 6x(25 - x^2)^{-5/2}$$

$$-\frac{15}{2}x^2(25 - x^2)^{-7/2}(-2x)$$

$$f'''(0) = 0$$

$$\frac{1}{\sqrt{5^2 - x^2}} = \frac{1}{5} + \frac{1}{250} x^2.$$

WORKED EXAMPLE 54

Using the expansion of $\dfrac{6}{x^2 - 9}$ derive the expansion of $\ln \dfrac{x-3}{x+3}$.

SOLUTION 54

$$\frac{6}{x^2 - 9} = 6 (x^2 - 9)^{-1} = 6 \left[x^2 (1 - 9/x^2)\right]^{-1} = 6 x^{-2} (1 - 9/x^2)^{-1}$$

$$= \frac{6}{x^2} \left[1 + (-1)\left(-\frac{9}{x^2}\right) + (-1)(-2)\left(-\frac{9}{x^2}\right)^2 \frac{1}{2!} + ...\right]$$

$$= \frac{6}{x^2} + \frac{54}{x^4} + \frac{486}{x^6} + ...$$

$$\int \frac{6}{x^2 - 9} \, dx = 6 \left[\frac{1}{2 \times 3} \quad \ln \frac{x-3}{x+3}\right]$$

$$= \ln \frac{x-3}{x+3} \quad \text{when} \mid x \mid > 3$$

$$\ln \frac{x-3}{x+3} = \int \left(\frac{6}{x^2} + \frac{54}{x^4} + \frac{486}{x^6} + ...\right) dx$$

$$= \int (6 x^{-2} + 54 x^{-4} + 486 x^{-6} + ...) \, dx$$

$$= \frac{6 x^{-1}}{-1} + \frac{54 x^{-3}}{-3} + \frac{486 x^{-5}}{-5} + ... = -\frac{6}{x} - \frac{18}{x^3} - \frac{486}{5} x^5 + ...$$

LEIBNITZ'S THEOREM

This theorem is a generalisation of the product rule in differentiation.

If n is a positive integer

$$(uv)_n = u_n v + n C_1 u_{n-1} v_1 + {}^{nC_r} u_{n-2} v_2 + \ldots + n Cr u_{n-r} v_r + \ldots + uv_n.$$

The product rule

$$\frac{d}{dx}(uv) = \frac{du}{dx} v + u \frac{dv}{dx}$$

which can be written as

$$(uv)_1 = u_1 v + u v_1$$

$$\frac{d^2}{dx^2}(uv) = \frac{d}{dx}\left(\frac{du}{dx} v + u \frac{dv}{dx}\right) = \frac{d^2u}{dx^2} v + \frac{du}{dx}\frac{dv}{dx}$$

$$+\frac{du}{dx}\frac{dv}{dx} + u \frac{d^2v}{dx^2} = u_2 v + 2 u_1 v_1 + u v_2$$

$$= u_2 v + 2 u_1 v_1 + u v_2$$

$$= u_2 v + {}^2C_1 u_1 v_1 + u v_2$$

$$\frac{d^3}{dx^3}(uv) = \frac{d}{dx}\left(\frac{d^2}{dx^2}(u\,v)\right) = \frac{d}{dx}(u_2 v + {}^2C_1 u_1 v_1 + u v_2)$$

$$= u_3 v + u_2 v_1 + {}^2C_1 u_2 v_1 + {}^2C_1 u_1 v_2 + u_1 v_2 + u v_3$$

$$= u_3 v + 3 u_2 v_1 + 3 u_1 v_2 + u v_3$$

$$= u_3 v + {}^3C_1 u_2 v_1 + {}^3C_2 u_1 v_2 + u v_3$$

$$\boxed{\frac{d^n}{dx^n}(uv)\, v = (u\,v)_n = u_n v + {}^{nC_1} u_{n-1} v_1 + {}^{nC_2} u_{n-2} v_2 + \ldots \\ + {}^{\cdot nC_r} u_{n-r} v_r + \ldots + u v_n.}$$

NUMERICAL METHODS FOR THE SOLUTION OF DIFFERENTIAL EQUATIONS POLYNOMIAL APPROXIMATIONS USING TAYLOR SERIES

The Taylor expansion of $f(x)$ about $x - x_0$ is given by

$$y = yo + \left(\frac{dy}{dx}\right)_0 (x - x_0) + \left(\frac{d^2y}{dx^2}\right)_0 \frac{(x - x_0)^2}{2!} + \left(\frac{d^3y}{dx^3}\right)_0 \frac{(x - x_0)^3}{3!} + \ldots$$

where y_0, $\left(\frac{dy}{dx}\right)_0$, $\left(\frac{d^2y}{dx^2}\right)_0$, $\left(\frac{d^3y}{dx^3}\right)_0$, $\ldots$ are the values at $x = x_0$.

This method is best illusrated by an example.

WORKED EXAMPLE 55

Find, as a series of ascending powers of x up to and including the term in x^5, an approximate solution to the differential equation $\frac{dy}{dx} = y e^{-x}$ where $y = 1$ when $x = 0$.

SOLUTION 55

$$\frac{dy}{dx} = y e^{-x}$$

differentiating with respect to x

$$\frac{d^2y}{dx^2} = \frac{dy}{dx} e^{-x} + y(-e^{-x})$$

$$\frac{d^2y}{dx^2} = \frac{dy}{dx} e^{-x} - y e^{-x}$$

differentiating with respect to x

$$\frac{d^3y}{dx^3} = \frac{d^2y}{dx^2} e^{-x} - \frac{dy}{dx} e^{-x} - \frac{dy}{dx} e^{-x} + y\, e^{-x}$$

$$\frac{d^3y}{dx^3} = \frac{d^2y}{dx^2} e^{-x} - 2\frac{dy}{dx} e^{-x} - \frac{dy}{dx} e^{-x} + y\, e^{-x}$$

differentiate again

$$\frac{d^4y}{dx^4} = \frac{d^3y}{dx^2} e^{-x} - \frac{d^2y}{dx^2} e^{-x} - 2\frac{d^2y}{dx^2} e^{-x} + 2\frac{dy}{dx} e^{-x} - \frac{d^2y}{dx^2} e^{-x}$$

$$+ \quad \frac{dy}{dx} e^{-x} + \frac{dy}{dx} e^{-x} - y\, e^{-x}$$

$$\frac{d^4y}{dx^4} = \frac{d^3y}{dx^3} e^{-x} - 4\frac{d^2y}{dx^2} e^{-x} + 4\frac{dy}{dx} e^{-x} - y\, e^{-x}$$

differentiate again

$$\frac{d^5y}{dx^5} = \frac{d^4y}{dx^4} e^{-x} - \frac{d^3y}{dx^3} e^{-x} - 4\frac{d^3y}{dx^3} e^{-x} + 4\frac{d^2y}{dx^2} e^{-x}$$

$$+ \quad 4\frac{d^2y}{dx^2} e^{-x} - 4\frac{dy}{dx} e^{-x} - \frac{dy}{dx} e^{-x} + y\, e^{-x}$$

$$\frac{d^5y}{dx^5} = \frac{d^4y}{dx^4} e^{-x} - 5\frac{d^3y}{dx^3} e^{-x} + 8\frac{d^2y}{dx^2} e^{-x} - 5\frac{dy}{dx} e^{-x} + y\, e^{-x}$$

$y = 1$ when $x = 0$

$$\left(\frac{dy}{dx}\right)_0 = y_0\, e^{-x_0}, \left(\frac{dy}{dx}\right)_0 = 1\, e^0 = 1$$

$$\left(\frac{d^2y}{dx^2}\right)_0 = \left(\frac{dy}{dx}\right)_0 e^{-x_0} - y_0\, e^{-x_0}, \left(\frac{d^2y}{dx^2}\right)_0 = (1)\,(1) - (1)\,(1) = 0$$

$$\left(\frac{d^3y}{dx^3}\right)_0 = \left(\frac{d^2y}{dx^2}\right)_0 e^{-x_0} - 3\left(\frac{dy}{dx}\right)_0 e^{-x_0} + y_0\, e^{-x_0}$$

$$\left(\frac{d^3y}{dx^3}\right) = (0)\,(1) - 3\,(1)\,(1) + (1)\,(1) = -2$$

$$\left(\frac{d^4y}{dx^4}\right)_0 = \left(\frac{d^3y}{dx^3}\right)_0 e^{-x_0} - 4\left(\frac{d^2y}{dx^2}\right)_0 e^{-x_0} + 4\left(\frac{dy}{dx}\right)_0 e^{-x_0} - y_0\, e^{-x_0}$$

$$= (-2)\,(1) - 4\,(0)\,(1) + 4\,(1)\,(1) - (1)\,(1) = 1$$

$$\left(\frac{d^5y}{dx^5}\right)_0 = \left(\frac{d^4y}{dx^4}\right)_0 e^{-x_0} - 5\left(\frac{d^3y}{dx^3}\right)_0 e^{-x_0} + 8\left(\frac{d^2y}{dx^2}\right)_0 e^{-x_0} - 5\left(\frac{dy}{dx}\right)_0 e^{-x_0} + y_0\, e^{-x_0}$$

$$= (1)\,(1) - 5\,(-2)\,(1) + 8\,(0)\,(1) - 5\,(1)\,(1) + (1)\,(1)$$

$$\left(\frac{d^5y}{dx^5}\right)_0 = 1 + 10 + 0 - 5 + 1 = 7$$

Applying

$$y \approx y_0 + \left(\frac{dy}{dx}\right)_0 (x - x_0) + \left(\frac{d^2y}{dx^2}\right)_0 \frac{(x - x_0)}{2!} + \dots$$

$$y \approx 1 + x\,(-2)\,\frac{x^3}{3!} + (1)\,\frac{x^4}{4!} + (7)\,\frac{x^5}{5!}$$

$$y \approx 1 + x - \frac{1}{3}\,x^3 + \frac{1}{24}\,x^4 + \frac{7}{120}\,x^5.$$

The exact solution of the differential equation

$$\frac{dy}{dx} = y\, e^{-x} \quad \text{where } y = 1 \text{ when } x = 0$$

$$\frac{dy}{dx} = e^{-x}\, dx$$

integrating both sides

$$\int \frac{dy}{dx} = \int e^{-x} \, dx$$

$\ln y = -e^{-x} + c$ $\qquad$ $\ln 1 = -e^0 + c$ $\qquad$ $0 = -1 + c$

$c = 1$ $\qquad\qquad$ $\ln y = -e^{-x} + 1$

$y = e^{-e^{-x} + 1}$

<table>
<tr><td colspan="2" align="center">**APPROXIMATE VALUES**</td><td align="center">**EXACT VALUES**</td></tr>
<tr><td>x</td><td>$y \approx 1 + x - \dfrac{1}{3} x^3 + \dfrac{1}{24} x^4 + \dfrac{7}{120} x^5$</td><td>$y = e^{1 - e^{-x}}$</td></tr>
<tr><td>0</td><td>1</td><td>1.</td></tr>
<tr><td>0.1</td><td>1.066709</td><td>1.0998377</td></tr>
<tr><td>0.5</td><td>1.462798</td><td>1.4821138</td></tr>
<tr><td>1.0</td><td>1.7666667</td><td>1.8815964</td></tr>
<tr><td>1.5</td><td>2.0289063</td><td>2.1746546</td></tr>
<tr><td>2.0</td><td>2.8666653</td><td>2.3742099</td></tr>
</table>

WORKED EXAMPLE 56

Find, as a series of ascending powers of x up to and including the term in x^4 an approximate solution to the differential equation $\dfrac{dy}{dx} = \dfrac{y^2 - 1}{x}$ where $y = 2$ when $x = 1$.

SOLUTION 56

$x \dfrac{dy}{dx} = y^2 - 1, \qquad x_0 \left(\dfrac{dy}{dx} \right)_0 = y_0^2 - 1, \qquad (1) \left(\dfrac{dy}{dx} \right)_0 = (2)^2 - 1$

$\left(\dfrac{dy}{dx} \right)_0 = 2.$

Differentiating with respect to x

$$\frac{dy}{dx} + x \frac{d^2y}{dx^2} = 2y \text{ repeating the differentiation } \frac{d^2y}{dx^2} + \frac{d^2y}{dx^2} + x \frac{d^3y}{dx^3} = 2 \frac{dy}{dx}$$

$$2\frac{d^2y}{dx^2} + x \frac{d^3y}{dx^3} = 2 \frac{dy}{dx}$$

Differentiating

$$2 \frac{d^3y}{dx^3} + \frac{d^3y}{dx^3} + x \frac{d^4y}{dx^4} = 2 \frac{d^2y}{dx^2} \qquad 3 \frac{d^3y}{dx^3} + x \frac{d^4y}{dx^4} = 2 \frac{d^2y}{dx^2}$$

$$\left(\frac{dy}{dx}\right)_0 + x_0\left(\frac{d^2y}{dx^2}\right)_0 = 2 y_0 \qquad 2 + (1)\left(\frac{d^2y}{dx^2}\right)_0 = 2 (2)$$

$$\left(\frac{d^2y}{dx^2}\right)_0 = 2 \qquad 2\left(\frac{d^2y}{dx^2}\right)_0 + {}^{x_0}\left(\frac{d^3y}{dx^3}\right)_0 = 2 \left(\frac{dy}{dx}\right)_0$$

$$2 (2) + (1)\left(\frac{d^3y}{dx^3}\right)_0 = 2 (1) \qquad \left(\frac{d^3y}{dx^3}\right)_0 = -2$$

$$3\left(\frac{d^3y}{dx^3}\right)_0 + x_0\left(\frac{d^4y}{dx^4}\right)_0 = 2\left(\frac{d^2y}{dx^2}\right)_0 \qquad 3 (-2) + (1)\left(\frac{d^4y}{dx^4}\right)_0 = 2 (2)$$

$$\left(\frac{d^4y}{dx^4}\right)_0 = 4 + 6 = 10 \qquad y \approx y_0 + \left(\frac{dy}{dx}\right)_0 (x - x_0) + \left(\frac{d^2y}{dx^2}\right)_0 \frac{(x - x_0)^2}{2!} +$$

$$y \approx 2 + (2) (x - 1) + (2) \frac{(x - 1)^2}{2} + (-2) \frac{(x - 1)^3}{6} + 10 \frac{(x - 1)^4}{24}$$

$$y \approx 2 + 2x - 2 + x^2 - 2x \quad \frac{x^3}{3} + 3 \frac{x^2}{3} - 3 \frac{x}{3} + \frac{1}{3}$$

$$+ \frac{5}{12} (x^4 - 4 x^3 + 6 x^2 - 4x + 1)$$

$$y \approx 1 + \frac{1}{3} + \frac{5}{12} + 2x - 2x - x - \frac{5}{3}x + x^2 + x^2 \qquad + \frac{5}{2}x^2 - \frac{1}{3}x^3 - \frac{5}{3}x^3 + \frac{5}{12}x^4$$

$$y \approx \frac{21}{12} - \frac{8}{3}x + \frac{9}{2}x^2 - 2x^3 + \frac{5}{12}x^4$$

$$\frac{dy}{y^2 - 1} = \frac{dx}{x}$$

$$\int \frac{dy}{(y - 1)(y + 1)} = \int \frac{dx}{x}$$

$$\frac{1}{(y - 1)(y + 1)} \equiv \frac{A}{y - 1} + \frac{B}{y + 1}$$

$$1 \equiv A(y + 1) + B(y - 1)$$

If $y = 1, A = \frac{1}{2}$; if $y = -1, B = -\frac{1}{2}$.

$$\frac{1}{(y - 1)(y + 1)} \equiv \frac{\frac{1}{2}}{y - 1} + - \frac{\frac{1}{2}}{y + 1}$$

$$\int \left[\frac{1}{2}(y - 1) - \frac{1}{2}(y + 1) \right] dy = \frac{1}{2} \ln |y - 1| - \frac{1}{2} |y + 1|$$

$$\frac{1}{2} \ln \frac{y - 1}{y + 1} = \ln x + \ln A \text{ where } x > 0$$

$$y > 1 \qquad y \neq -1$$

when $y = 2, x = 1, \quad \frac{1}{2} \ln \frac{1}{3} = \ln 1 + \ln A$

$$\frac{1}{2} \ln \frac{y - 1}{y + 1} = \ln x + \frac{1}{2} \ln \frac{1}{3}$$

$$\ln \sqrt{\frac{y - 1}{y + 1}} = \ln x \sqrt{\frac{1}{3}} \qquad\qquad \frac{y - 1}{y + 1} = \frac{1}{3}x^2$$

$$3y - 3 = yx^2 + x^2 \qquad 3y - yx^2 = x^2 + 3 \qquad y = \frac{x^2 + 3}{3 - x^2}$$

$$y \approx \frac{21}{12} - \frac{8}{3}x + \frac{9}{2}x^2 - 2x^3 + \frac{5}{12}x^4 \qquad y = \frac{x^2 + 3}{3 - x^2}$$

	Approximate values	**Exact values**
x		
1	2	2
0.5	1.3180416	1.1818182

WORKED EXAMPLE 57

Find the solution, in ascending powers of x up to and including the term x^3, of the differential equation

$$\frac{d^2y}{dx^2} + (x + 1)\frac{dy}{dx} + y = 0$$

giving that, when $x = 0$, $y = 1$ and $\dfrac{dy}{dx} = 2$.

SOLUTION 57

$$\frac{d^2y}{dx^2} + (x + 1)\frac{dy}{dx} + y = 0 \;\ldots(1) \qquad \left(\frac{d^2y}{dx^2}\right)_0 + (x_o + 1)\left(\frac{dy}{dx}\right)_0 + y_o = 0$$

$$x_0 = 0, \; y_0 = 1, \left(\frac{dy}{dx}\right)_0 = 2$$

$$\left(\frac{d^2y}{dx^2}\right)_0 + (2) + 1 = 0 \qquad\qquad \left(\frac{d^2y}{dx^2}\right)_0 = -3$$

differentiating equation (1) with respect to x

$$\frac{d^3y}{dx^3} + \frac{dy}{dx} + (x + 1)\frac{d^2y}{dx^2} + \frac{dy}{dx} = 0$$

$$\left(\frac{d^3y}{dx^3}\right)_0 + \left(\frac{dy}{dx_0}\right) + (x_o + 1)\left(\frac{d^2y}{dx^2}\right)_0 + \left(\frac{dy}{dx}\right)_0 = 0$$

$$\left(\frac{d^3y}{dx^3}\right)_0 + 2 + (0 + 1)(-3) + (2) = 0 \qquad \left(\frac{d^3y}{dx^3}\right)_0 = -1$$

$$y \approx y_0 + \left(\frac{dy}{dx}\right)_0 (x - x_0) + \left(\frac{d^2y}{dx^2}\right)_0 \frac{(x - x_0)^2}{2!} + \dots$$

$$y \approx 1 + 2x + (-3)\frac{x^2}{2} + (-1)\frac{x^3}{6}$$

$$y \approx 1 + 2x - \frac{3}{2}x^2 - \frac{1}{6}x^3$$

EXERCISES 11

1. If $\cos x = 1 - \dfrac{x^2}{2!} + \dfrac{x^4}{4!} - \dfrac{x^6}{6!} + \ldots$ find the power series for $\sin x$.

2. Obtain the first three terms of the expansion $\sin^{-1} x$ using the Maclaurin's theorem.

3. If $\tan^{-1} = a_0 + a_1 x + a_2 x^2 + a_3 x^3 + a_4 x^4$ determine the coefficients a_0, a_1, a_2, a_3, a_4.

4. Determine the expansion for $\dfrac{1}{1 + x^2}$ using :-

   (a)   the binomial theorem
   (b)   the Maclaurin's theorem.

5. Using the expansion of $\dfrac{1}{25 - x^2}$, derive the expansion of $\ln \dfrac{5 + x}{5 - x}$, $\mid x$

6. Using the expansion of $\dfrac{1}{(x^2 + 4)^{1/2}}$, derive the expansion of $\sinh^{-1} \dfrac{x}{2}$.

7. Using the expansion of $\dfrac{1}{\sqrt{1 - x^2}}$, derive the expansion of $\sin^{-1} x \left(\mid x \mid < 1 \right)$.

8. Use Maclaurin's theorem to obtain the following expansions:-

   (i) $\sinh^2 x$ (ii) $\log_e (1 - x^2)$ (iii) $\sin^2 x$ (iv) $\cos^2 x$.

9. Use Taylor's theorem to obtain approximate values of the following:-

   (i) $\tan 45° 2'$ (ii) $\log_e 1.001$.

10. Determine the expansion of $\log_e (1 + x + x^2)$.

   Hint use $1 - x^3 = (1 - x)(1 + x + x^2)$ provided that $-1 \leq x < 1$. Hence determine $\ln 1.0204$.

11. Determine the expansions:-

(i) $\log_e (1 - 4x)$ (ii) $\log_e (3 + 5x)$ (iii) $\log_e (2 + 7x)$.

State in each case the range of values of x for which the expansions are valid.

12. Use the expansion of $\tan^{-1} x$ to deduce the result

$$\pi = 4\left(1 - \frac{1}{3} + \frac{1}{5} - \frac{1}{7} + \ldots\right).$$

13. Use the expansion of $\sin^{-1} x$ to deduce the result

$$\pi = 6\left(\frac{1}{2} + \frac{1}{2 \times 3 \times 2^3} + \frac{1 \times 3}{2 \times 4 \times 5 \times 2^5} + \ldots\right)$$

14. Find the first four terms in the expansion of $\cos^{-1} x$ determine the value of π to three significant figures. You may use the expansion of $\cos^{-1} x$ or derive it by Maclaurin's theorem.

15. By means of the Taylor's series method, derive the solution, as a power series of ascending powers of x as far as the term in x^6, of the differential

equation $\dfrac{d^2y}{dx^2} - 5x\dfrac{dy}{dx} + 2y = 0$, given that $y = 2$, $\dfrac{dy}{dx} = 1$ when $x = 0$.

16. Use Maclaurin's theorem to obtain the first three non-zero terms of the series expansions for $\sinh x$. Differentiate this to obtain the series for $\cosh x$. Hence or otherwise, obtain series expansions in terms of x up to and including the term in x^3 for:-

(i) $e^{-\cosh x}$ (ii) $e^{-\sinh x}$ (iii) $\coth x$.

17. If $y = \tan x$, find the first six derivatives using the notation $y_r = d^r y / dx^r$.

18. If $y = \ln \sin x$, find the first four derivatives. Obtain the Maclaurin's expansion of y in terms of x up to and including the term in x^5.

19. State the expansion of $y = \sinh nx$ as a series in ascending powers of x up to and including the term in n^5.

SOLUTIONS 1

1. $$y = ax^n \qquad \frac{dy}{dx} = anx^{n-1}$$

2.   (i) $y = 3$, when x increases by δx then y increases by δy

              $y + \delta y = 3$, subtracting the two equations $\delta y = 0$, dividing each

              term by δx $\frac{\delta y}{\delta x}$, when δx tends to zero then $\frac{\delta y}{\delta x}$ tends to

              $\frac{dy}{dx}$, and therefore

$$\boxed{\frac{dy}{dx} = 0}$$

   (ii) $y = x$, $y + \delta y = x + \delta x$ subtracting the two equations $\delta y = \delta x$,

              dividing by δx each term, we have, $\frac{\delta y}{\delta x} = 1$ as $\delta x \to 0$ then $\frac{\delta y}{\delta x} \to \frac{dy}{dx}$

$$\boxed{\frac{dy}{dx} = 1}$$

   (iii) $y = -x^2 + 1$, $y + \delta y = -(x + \delta x)^2 + 1$, subtracting

              $\delta y = -(x + \delta x^2 + 1 + x^2 - 1 = -x^2 - 2x\,\delta x - \delta x^2 + 1 + x^2 - 1$

              $\delta y = -2x - \delta x - \delta x^2$, dividing each term by

              $\frac{\delta y}{\delta x} = -2x - \delta x$, as $\delta x \to 0$, $\frac{\delta y}{\delta x} \to \frac{dy}{dx}$

$$\boxed{\frac{dy}{dx} = -2x}$$

   (iv) $y = 2x^3$, $y + \delta y = 2(x + \delta x)^3 = 2(x^3 + 3x^2\,\delta x + 3x\,\delta x^2 + \delta x^3)$

              subtracting the two equations $\delta y = 6x^2\,\delta x + 6x\,\delta x + 2\,\delta x^2$,

              dividing each term by δx $\frac{\delta y}{\delta x} = 6x^2 + 6x\,\delta x + 2\,\delta x^2$, as $\delta x \to 0$, $\dfrac{\delta y}{\delta x} \to \dfrac{dy}{dx}$ $\dfrac{dy}{dx} = 6\,x^2.$

(v) $y = 5x - \dfrac{3}{x} + \dfrac{1}{x^2}$, $y + \delta y = 5\,(x + \delta x) - \dfrac{3}{(x + \delta x)} + \dfrac{1}{(x + \delta x)^2}$

subtracting the two equations

$$\delta y = 5\,\delta x - \dfrac{3}{x - \delta x} + \dfrac{3}{x} + \dfrac{1}{x + \delta x^2} - \dfrac{1}{x^2}$$

$$= 5\,\delta x + \dfrac{3\,(x + \delta x) - 3x}{x\,(x + \delta x)} + \dfrac{x^2 - (x + (\delta x)^2}{x^2(x + \delta x^2}$$

$$= 5\,\delta x + \dfrac{3x + 3\,\delta x - 3x}{x\,(x + \delta x)} + \dfrac{x^2 - x^2 - 2x\delta x - \delta x^2}{x^2(x + \delta x)^2}$$

$$= 5\,\delta x + \dfrac{3\,\delta x}{x\,(x + \delta x)} - \dfrac{2x\,\delta x + \delta x^2}{x^2\,(x + \delta x)^2} \qquad \text{dividing each term by } \delta x.$$

$$\dfrac{\delta y}{\delta x} = 5 + \dfrac{3}{x\,(x + \delta x)} - \dfrac{2x + \delta x}{x^2\,(x + \delta x)^2} \qquad \text{as } \delta x \to 0, \quad \dfrac{\delta y}{\delta x} \to \dfrac{dy}{dx}$$

$$\dfrac{dy}{dx} = 5 + \dfrac{3}{x^2} - \dfrac{2x}{x^4} = 5 + \dfrac{3}{x^2} - \dfrac{2}{x^3} \qquad \dfrac{dy}{dx} = 5 + \dfrac{3}{x^2} - \dfrac{2}{x^3}.$$

3.   (i) $y = 3x$, $dy/dx = 3$

   (ii) $y = 5$, $y = 6\,x^0$, $\dfrac{dy}{dx} = 5 \times 0\,x^{-1} = 0$

   (iii) $y = \dfrac{3}{x} = 3\,x^{-1}$, $\dfrac{dy}{dx} = -3\,x^{-2} = -\dfrac{3}{x^2}$

   (iv) $y = -x^2 - x^3 - x^4$, $\dfrac{dy}{dx} = -2x - 3\,x^2 - 4x^3$

   (v) $y = \dfrac{3}{\sqrt{x}} = 3\,x^{-1/2}$, $\dfrac{dy}{dx} = -\dfrac{3}{2}\,x^{-3/2} = -\dfrac{3}{2\,x^{3/2}}$

   (vi) $y = \dfrac{1}{x} + \dfrac{4}{x^2} - \dfrac{3}{x^3} = x^{-1} + 4\,x^{-2} - 3\,x^{-3}$

   $$\dfrac{dy}{dx} = -x^{-2} - 8\,x^{-3} + 9\,x^{-4} = -\dfrac{1}{x^2} - \dfrac{8}{x^3} + \dfrac{9}{x^4}$$

(vii) $\quad x = 3\,t^2 - 5t, \dfrac{dx}{dt} = 6t - 5$

(viii) $\quad \Theta = 3\,t^2 - 5t, \dfrac{d\Theta}{dt} = 6t - 5$

(ix) $\quad r = \dfrac{1}{t} + t - t^2, \dfrac{dr}{dt} = -\dfrac{1}{t^2} + 1 - 2t$

(x) $\quad Z = 5\,y^2 - 5\,y^3 - 7\,y^5, \dfrac{dZ}{dy} = 10y - 15\,y^2 - 35\,y^4.$

4. (i) $\quad y = (x^2 + 3)\,(x^2 - 5), \dfrac{dy}{dx} = 2x.\,(x^2 - 5) + (x^2 + 3)\,2x$

$\qquad \dfrac{dy}{dx} = 2\,x^3 - 10x + 2\,x^3 + 6x = 4\,x^3 - 4x.$

(ii) $\quad t = (x + 1)\,(x^3 - a), \dfrac{dt}{dx} = 1.\,(x^3 - 9) + (x + 1)\,3x^2 = x^3 - 9 + 3\,x^3$

$\qquad \dfrac{dt}{dx} = 4\,x^3 + 3\,x^2 - 9.$

(iii) $\quad \Theta = (3\,t^3 - 5)\,t^5, \dfrac{d\Theta}{dt} = 9\,t^2.\,t^5 + (3\,t^3 - 5)\,5\,t^4 = 9\,t^7 + 15\,t^7 - 25$

$\qquad \dfrac{d\Theta}{dt} = 24\,t^7 - 25\,t^4.$

(iv) $\quad y = 3\,(x^{27} - 3), \dfrac{dy}{dx} = 0.\,(x^{27} - 3) + 3 \times 27\,x^{26} = 81\,x^{26}$

(v) $\quad y = x\,(x^2 - 1)\,(x^3 - 2),$

$\qquad \dfrac{dy}{dx} = 1.\,(x^2 - 1)\,(x^3 - 2) + x.2x.\,(x^3 - 2) + x\,(x^2 - 1)\,3\,x^2$

$\qquad \dfrac{dy}{dx} = (x^2 - 1)\,(x^3 - 2) + 2\,x^2\,(x^3 - 2) + 3\,x^3\,(x^2 - 1)$

5. (i) $\quad y = \dfrac{x^2}{x^5 - 1}, \dfrac{dy}{dx} = \dfrac{2x\,(x^5 - 1) - x^2.\,5\,x^4}{(x^5 - 1)^2}$

$$\frac{dy}{dx} = \frac{2\,x^6 - 2x - 5\,x^6}{x^5 - 1)^2} = -\frac{3\,x^6 + 2x}{(x^5 - 1)^2} = -\frac{x\,(3\,x^5 + 2)}{(x^5 - 1)^2}$$

(ii) $\quad Z = \dfrac{t^3 - 1}{t^3 + 1}, \dfrac{dZ}{dt} = \dfrac{3\,t^2\,(t^3 - 1) - (t^3 - 1)\,3\,t^2}{t^3 + 1)^2}$

$$\frac{dZ}{dt} = \frac{3\,t^5 + 3\,t^2 - 3\,t^5 + 3\,t^2}{(t^3 + 1)^2} = \frac{6\,t^2}{(t^3 + 1)^2}$$

(iii) $\quad \Theta = 3\dfrac{r}{r^4 - 1}, \dfrac{d\Theta}{dr} = \dfrac{3\,(r^4 - 1) - 3r.\,4\,r^3}{(r^4 - 1)^2}$

$$\frac{d\Theta}{dr} = \frac{3\,r^4 - 3 - 12\,r^4}{(r^4 - 1)^2} = \frac{-3 - 9r^4}{(r^4 - 1)^2} = -\frac{3\,(1 + 3\,r^4)}{(r^4 - 1)^2}$$

(iv) $\quad y = \dfrac{x^4 - 3\,x^3 + 2\,x^2}{5x} = \dfrac{1}{5}\,(x^3 - 3\,x^2 + 2x)$

$$\frac{dy}{dx} = \frac{3}{5}\,x^2 - \frac{6}{5}\,x + \frac{2}{5}$$

(v) $\quad y = \dfrac{x - 1}{x + 2}, \dfrac{dy}{dx} = \dfrac{1.\,(x + 2) - (x - 1).\,1}{(x + 2)^2} = \dfrac{3}{(x + 2)^2}$

$$\frac{dy}{dx} = \frac{3}{(x + 2)^2}.$$

6. (i) $\quad y = \sqrt{3x + 1} = (3x + 1)^{1/2} \quad$ let $u = 3x + 1, \dfrac{du}{dx} = 3$

$\quad y = u^{1/2}, \dfrac{dy}{du} = \dfrac{1}{2}\,u^{-1/2}, \dfrac{dy}{dx} = \dfrac{3}{2}\,(3x + 1)^{-1/2}$

$$\frac{dy}{dx} = \frac{3}{2\,(3x + 1)^{1/2}}$$

(ii) $\quad y = \dfrac{1}{\sqrt{x - 1}} = \dfrac{1}{(x - 1)^{1/2}} = (x - 1)^{-1/2} \quad$ Let $u = x - 1, \dfrac{du}{dx} = 1$

$\quad y = u^{-1/2}, \dfrac{dy}{du} = -\dfrac{1}{2}\,u^{-3/2}, \dfrac{dy}{dx} = -\dfrac{1}{2\,(x - 1)^{3/2}}$

(iii) $\quad y = (x^3 - 1)^3, \text{ let } u = x^3 - 1, \dfrac{du}{dx} = 3\,x^2$

$$y = u^3, \frac{dy}{dx} = 3\,u^2 = 3\,(x^3 - 1)^2$$

$$\frac{dy}{dx} = \frac{dy}{du} \cdot \frac{du}{dx} = 3\,(x^3 - 1)^2.\ 3\,x^2 = 9\,x^2 = 9\,x^2\,(x^3 - 1)^2$$

$$\frac{dy}{dx} = 9\,x^2\,(x^3 - 1)^2.$$

(iv) $\quad y = (x^2 - 1)^3 \cdot (x^3 + 1)^4 \quad$ let $u = (x^2 - 1)^3$, $v = (x^3 + 1)^4$

$$u = (x^2 - 1)^3 \quad \text{let } w = x^2 - 1,\ \frac{dw}{dx} = 2x$$

$$u = w^3,\ \frac{du}{dw} = 3\,w^2 = 3\,(x^2 - 1)^2,\ \frac{du}{dx} = 2x.\ 3\,(x^2 - 1)^2 = 6x\,(x^2 - 1)^2$$

$$v = (x^3 + 1)^4 \quad \text{let } z = x^3 + 1,\ \frac{dz}{dx} = 3\,x^2$$

$$v = z^4,\ \frac{dv}{dz}\ = 4\,z^3 = 4\,(x^3 + 1)^3,\ \frac{dv}{dx} = 3\,x^2.\ 4\,(x^3 + 1)^3$$

$$= 12\,x^2\,(x^3 + 1)^3$$

$$\frac{dy}{dx} = 6x\,(x^2 - 1)^2\,(x^3 + 1)^4 + (x^2 - 1)^3.\ 12\,x^2\,(x^3 + 1)^3$$

$$= 6x\,(x^2 - 1)^2\,(x^3 + 1)\left(x^3 + 1 + x^2 - 1 + 2\right)$$

$$= 6x\,(x^2 - 1)^2\,(x^3 + 1)^3\,(x^3 + x^2 + 2)$$

(v) $\quad y = (5\,x^2 - 7)^{1/3} \quad$ let $u = 5x^2 - 7\ \ \dfrac{du}{dx} = 10x$

$$y = u^{1/3},\ \frac{dy}{du} = \frac{1}{3}\,u^{-2/3} = \frac{1}{3}\,(5\,x^2 - 7)^{-2/3}$$

$$\frac{dy}{dx} = \frac{10}{3}\,x\,(5\,x^2 - 7)^{-2/3}.$$

140

7.   An explicit function is an expression in which one variable exists on one side of the equation and another variable exists on the other side.

such as $y = x$, $y = \dfrac{x^2 + 1}{x - 1}$, $z = 3t$

where y is expressed in terms of x of z is expressed in terms of t the independent variable is x and y is the dependent variable, or the independent variable is t and z is the dependent variable.

$xy = 3$ may be expressed explicitly if y is expressed in terms of x $y = \dfrac{3}{x}$.

$x^2 y + y^2 = 1$, however, cannot be easily expressed explicitly, this expression is termed as an implicit function.   If we consider $x^2 y + y^2 = 1$ as a quadratic in y,

then $y^2 + xy - 1 = 0$, the solution will be $y = - \dfrac{x \pm \sqrt{x^2 + 4}}{2}$

so y can be expressed explicitly with some difficulty, to find dy/dx from this expression is rather difficult, so differentiation of implicit functions can be mastered.

8.   (a)   (i) $xy = c^2$.

differentiating with respect to x, treat the lefthand side as a product.

$$\frac{d}{dx}(x) \cdot y + x \frac{d}{dx}(y) = \frac{d}{dx}(c^2) \qquad 1.y + x\frac{dy}{dx} = 0$$

therefore $\dfrac{dy}{dx} = -y/x \;\ldots(1)$

If dy/dx is required in terms of x then substitute $y = c^2/x$ in (1)

$$\frac{dy}{dx} = -\frac{c^2}{x^2}$$

(ii) $\dfrac{x^2}{a^2} - \dfrac{y^2}{b^2} = 1$

differentiating with respect to x

$$\frac{d}{dx}\left(\frac{x^2}{a^2}\right) - \frac{d}{dx}\left(\frac{y^2}{b^2}\right) = \frac{d}{dx}(1)$$

141

$$\frac{2x}{a^2}\frac{dx}{dx} - \frac{2y}{b^2}\frac{dy}{dx} = 0 \qquad -\frac{2y}{b^2}\frac{dy}{dx} = -\frac{2x}{a^2}$$

$$\frac{dy}{dx} = \frac{b^2}{a^2}\frac{x}{y}$$

is not always easy to express $\dfrac{dy}{dx}$ in terms of x.

(iii) $\qquad \dfrac{y^2}{a^2} - \dfrac{x^2}{b^2} = 1 \quad$ differentiating with respect to x

$$\frac{2y}{a^2}\frac{dy}{dx} - \frac{2x}{b^2} = 0, \frac{dy}{dx} = \frac{x}{y}\frac{a^2}{b^2}$$

(iv) $\qquad x^2 + y^2 = r^2$ differentiating with respect to x $\quad 2x + 2y\dfrac{dy}{dx} = 0$

$$\frac{dy}{dx} = -\frac{x}{y}$$

(v) $\qquad \dfrac{x^2}{a^2} + \dfrac{y^2}{b^2} = 1 \qquad\qquad \dfrac{2x}{a^2} + \dfrac{2y}{b^2}\dfrac{dy}{dx} = 0, \dfrac{dy}{dx} = -\dfrac{x}{y}\dfrac{a^2}{b^2}$

(vi) $\qquad xy + y^2 = 5$ differentiating with respect to x $\quad 1.y + x\dfrac{dy}{dx} + 2y\dfrac{dy}{dx} =$

$$\frac{dy}{dx}(x + 2y) = -y \qquad \frac{dy}{dx} = -\frac{y}{x + 2y}$$

(vii) $\qquad x^2 + y^2 - 3xy + \delta y = 0$ differentiating with respect to x

$$2x + 2y\frac{dy}{dx} - 3y - 3x\frac{dy}{dx} + 5\frac{dy}{dx} = 0$$
$$\frac{dy}{dx}(2y - 3x + 5) = 3y - 2x$$

$$\frac{dy}{dx} = \frac{3y - 2x}{2y - 3x + 5}$$

(viii) $\qquad x^2 + y^2 + 2gx + 2fy + c = 0$ differentiating with respect to x

$$2x + 2y\frac{dy}{dx} + 2g + 2f\frac{dy}{dx} = 0$$

$$\frac{dy}{dx} = -\frac{x+g}{y+f}$$

(ix) $y^2 = 4ax$ differentiating with respect to x $2y\,\dfrac{dy}{dx} = 4a$

$$\frac{dy}{dx} = \frac{2a}{y}$$

(x) $x^2 = -5y$ differentiating with respect to $2x = -5\,\dfrac{dy}{dx}$

$$\frac{dy}{dx} = -\frac{2}{5}\,x$$

(b)   (i) $xy = c^2$ (ii) $\dfrac{x^2}{a^2} - \dfrac{y^2}{b^2} = 1$

$$\frac{dx}{dy}\cdot y + x.1 = 0 \qquad\qquad \frac{2x}{a^2}\frac{dx}{dy} - \frac{2y}{b^2} = 0$$

$$\frac{dx}{dy} = -\frac{x}{y} \qquad\qquad\qquad \frac{dx}{dy} = \frac{y}{x}\frac{a^2}{b^2}$$

(iii) $\dfrac{y^2}{a^2} - \dfrac{x^2}{b^2} = 1$ (iv) $x^2 + y^2 = r^2$

$$\frac{2y}{a^2} - \frac{2x}{b^2}\frac{dx}{dy} = 0 \qquad\qquad 2x\,\frac{dx}{dy} + 2y = 0$$

$$\frac{dx}{dy} = \frac{y}{x}\frac{a^2}{b^2} \qquad\qquad\qquad \frac{dx}{dy} = -\frac{y}{x}$$

(v) $\dfrac{x^2}{a^2} + \dfrac{y^2}{b^2} = 1$ (vi) $xy + y^2 = 5$

$$\frac{2x}{a^2}\frac{dx}{dy} + \frac{2y}{b^2} = 0 \qquad\qquad \frac{dx}{dy}\cdot y + x.1 + 2y = 0$$

$$\frac{dx}{dy} = -\frac{y}{x}\frac{a^2}{b^2} \qquad\qquad \frac{dx}{dy} = -\frac{(x+2y)}{y}$$

(vii) $x^2 + y^2 - 3xy + 5y = 0$ (viii) $x^2 + y^2 + 2gx + 2fy + c = 0$

$$2x \frac{dx}{dy} + 2y - 3 \frac{dx}{dy} y - 3x + 5 = 0 \qquad\qquad 2x \frac{dx}{dy} + 2y + 2g \frac{dx}{dy} + 2f = 0$$

$$\frac{dx}{dy} = \frac{3x - 2y - 5}{2x - 3y} \qquad\qquad\qquad \frac{dx}{dy} = - \frac{f + y}{x + g}$$

(ix) $\quad y^2 = 4ax$ $\qquad\qquad\qquad\qquad$ (x) $\qquad x^2 = -5y$

$$2y = 4a \frac{dx}{dy} \qquad\qquad\qquad\qquad\qquad 2x \frac{dx}{dy} = -5$$

$$\frac{dx}{dy} = \frac{y}{2a} \qquad\qquad\qquad\qquad\qquad\quad \frac{dx}{dy} = - \frac{5}{2x}$$

9. $\quad$ (i) $\qquad y = 3x^2 - 5x - 7$ $\qquad$ (ii) $\qquad y = \frac{x}{x - 1}$

$$\frac{dy}{dx} = 6x - 5 \qquad\qquad\qquad \frac{dy}{dx} = \frac{x - 1 - x}{(x - 1)^2} = - \frac{1}{(x - 1)^2}$$

at $x = -1$

$$\frac{dy}{dx} = -6 - 5 = -11 \qquad\qquad \text{at } x = 2$$

$$\frac{dy}{dx} = - \frac{1}{1^2} = -1$$

(iii) $\qquad y = x (x^3 - 1)(x^2 + 1)$

$$\frac{dy}{dx} = (x^3 - 1)(x^2 + 1) + x.3x^2 (x^2 + 1) + (x^3 - 1).2x$$

at $x = 0$

$$\frac{dy}{dx} = (-1)(1) = -1$$

(iv) $\qquad y = \frac{5}{x} = 5x^{-1}, \frac{dy}{dx} = -5x^{-2}$ $\quad$ at $x = 5$, $\quad \frac{dy}{dx} = - \frac{5}{5^2} = - \frac{1}{5}.$

(v) $y = -x^{-2/3}$, $\dfrac{dy}{dx} = \dfrac{2}{3} x^{-5/3}$ at $x = 1$, $\dfrac{dy}{dx} = \dfrac{2}{3}$

(vi) $y^2 = x$, $2y \dfrac{dy}{dx} = 1$, $\dfrac{dy}{dx} = \dfrac{1}{2y}$ at $x = 4$, $y^2 = 4$, $y = \pm 2$

$\dfrac{dy}{dx} = \pm \dfrac{1}{4}$

(vii) $x^2 = -4y$ $2x = -4 \dfrac{dy}{dx}$

$\dfrac{dy}{dx} = -\dfrac{x}{2}$ when $x = -2$, $\dfrac{dy}{dx} = 1$

(viii) $y = \dfrac{1}{x} + \dfrac{2}{x^2} + \dfrac{3}{x^3} = x^{-1} + 2x^{-2} + 3x^{-3}$ $\dfrac{dy}{dx} = -x^{-2} - 4x^{-3} - 9x^{-4}$

at $x = -1$, $\dfrac{dy}{dx} = -1 + 4 - 9 = -6$

(ix) $x^2 + y^2 = 4$, $2x + 2y \dfrac{dy}{dx} = 0$ at $x = y^2 = 4 - 1 = 3$, $y = \pm \sqrt{3}$

$\dfrac{dy}{dx} = -\dfrac{x}{y} = \mp \dfrac{1}{\sqrt{3}}$

(x) $xy - x^2 + y^2 - 1 = 0$ $1 y + x \dfrac{dy}{dx} - 2x + 2y \dfrac{dy}{dx} = 0$

when $x = 0$, $y = \pm 1$

$\dfrac{dy}{dx} = \dfrac{2x - y}{2y + x} = -\dfrac{1}{2}.$

10. (i) $y = anx^{n-1}$ $\dfrac{dy}{dx} = an(n-1)x^{n-2}$

(ii) $y = \dfrac{1}{\sqrt{x}} + \sqrt{x} + \sqrt[3]{x^2} = x^{-1/2} + x^{1/2} + x^{2/3}$

$\dfrac{dy}{dx} = -\dfrac{1}{2} x^{-3/2} + \dfrac{1}{2} x^{-1/2} + \dfrac{2}{3} x^{-1/3}$

(iii) $y = \sqrt{x^2 - 1} \sqrt{x^2 + 1} = (x^2 - 1)^{1/2} (x^2 + 1)^{1/2} = (x^4 - 1)^{1/2}$

Let $u = x^4 - 1$ $\dfrac{dy}{dx} = 4x^3$ $y = u^{1/2}, \dfrac{dy}{du} = \dfrac{1}{2} u^{-1/2}$

$$\frac{dy}{dx} = 2x^3 (x^4 - 1)^{-1/2} = \frac{2x^3}{(x^4 - 1)^{1/2}}$$

(iv) $\quad y = \dfrac{x^2 - 1}{x + 1} \qquad \dfrac{dy}{dx} = \dfrac{2x \cdot (x + 1) - (x^2 - 1)}{(x + 1)^2} = \dfrac{2x^2 + 2x - x^2 + 1}{(x + 1)^2}$

$$\frac{dy}{dx} = \frac{x^2 + 2x + 1}{(x + 1)^2} = \frac{(x + 1)^2}{(x + 1)^2} = 1$$

(v) $\quad y = \dfrac{x^3 - 1}{x^3 + 2} \qquad \dfrac{dy}{dx} = \dfrac{3x^2 (x^3 + 2) - (x^3 - 1) 3x^2}{(x^3 + 2)^2}$

$$= \frac{3x^5 + 6x^2 - 3x^5 + 3x^2}{(x^3 + 2)^2} = \frac{9x^2}{(x^3 + 2)^2}$$

(vi) $\quad y = (1 - 3x)^{1/5} \qquad$ let $u = 1 - 3x \qquad \dfrac{dy}{dx} = -3$

$$y = u^{1/5} \qquad \frac{dy}{du} = \frac{1}{5} u^{-4/5} \qquad \frac{dy}{dx} = -\frac{3}{5(1 - 3x)^{4/5}}$$

(vii) $\quad y = x^2 \sqrt{x - 1} = x^2 (x - 1)^{1/2}$

$$\frac{dy}{dx} = 2x \cdot (x - 1)^{1/2} + x^2 \frac{1}{2} (x - 1)^{-1/2}$$

(viii) $\quad xy = x^2 + y^2$

$$1 \cdot y + x \frac{dy}{dx} = 2x + 2y \frac{dy}{dx}, \frac{dy}{dx} (x - 2y) = 2x - y \qquad \frac{dy}{dx} = \frac{2x - y}{x - 2y}$$

(ix) $\quad 3x^2 + 3y = x - y - 5 \qquad\qquad 6x + 6y \dfrac{dy}{dx} = 1 - \dfrac{dy}{dx} - 5$

$$\frac{dy}{dx} (6x + 1) = 1 - 5 - 6x \qquad \frac{dy}{dx} = \frac{1 - 5 - 6x}{6x + 1} = -\frac{4 + 6x}{6x + 1}$$

(x) $\quad 3x^2 + 5y^2 = 25 \quad 6x + 10y \dfrac{dy}{dx} = 0 \qquad\qquad \dfrac{dy}{dx} = -\dfrac{3x}{5y}.$

11.   (i) $y = (2x + 5)^4$

$$\frac{dy}{dx} = 8\,(2x + 5)^3$$

(ii) $y = (x - 1)^{-4}$

$$\frac{dy}{dx} = -4\,(x - 1)^{-5}$$

$$= -4/(x - 1)^5$$

(iii) $y = (1 + 3x)^{1/2}$

$$\frac{dy}{dx} = \frac{1}{2}\,(1 + 3x)^{-1/2}\,(3)$$

$$= \frac{3}{2}\,(1 + 3x)^{1/2}$$

(iv) $y = (1 + 2 + 3\,x^2)^5$

$$\frac{dy}{dx} = 5\,(1 + 2x + 3\,x^2)^4\,(2 + 6x)$$

(v) $y = (5x - 7)^{1/2}$

$$\frac{dy}{dx} = \frac{1}{2}\,(5x - 7)^{-1/2}\,(5)$$

$$= \frac{5}{2\,(5x - 7)^{1/2}}$$

(x) $y = \dfrac{x^3 - 2\,x^2 - 7x + 1}{x^2 - 1}$

(vi) $y = (3x - 2)^{-1/2}$

$$\frac{dy}{dx} = -\frac{1}{2}\,(3x - 2)^{-3/2}\,(3)$$

$$= -\frac{3}{2}\,(3x - 2)^{3/2}$$

(vii) $y = (1 + 7x)^{1/2}$

$$\frac{dy}{dx} = \frac{1}{2}\,(1 + 7x)^{-1/2}\,(7)$$

$$= \frac{7}{2}\,(1 + 7x)^{1/2}$$

(viii) $y = \dfrac{1}{\sqrt{1 + 4\,x^2}} = (1 + 4\,x^2)^{-1/2}$

$$\frac{dy}{dx} = -\frac{1}{2}\,(1 + 4\,x^2)^{-3/2}\,(8x)$$

$$= \frac{4x}{(1 + 4\,x^2)^{3/2}}$$

(ix) $y = (5\,x^2 + 7x - 3)^2$

$$\frac{dy}{dx} = 2\,(5\,x^2 + 7x - 3)\,(10x + 7)$$

$$\frac{dy}{dx} = \frac{3x^4 - 4x^3 - 7x^2 - 3x^2 + 4x + 7 - 2x^4 + 4x^3 + 14x^2 - 2x}{(x^2 - 1)^2}$$

$$\frac{dy}{dx} = \frac{x^4 + 4x^2 + 2x + 7}{(x^2 - 1)^2}$$

12. (i) $\sqrt{x} + \sqrt{y} = \sqrt{3}$ or $x^{1/2} + y^{1/2} = \sqrt{3}$, differentiating w.r.t.x

$$\frac{1}{2} x^{-1/2} + \frac{1}{2} y^{-1/2} \frac{dy}{dx} = 0$$

$$\frac{dy}{dx} = -\frac{1}{2} x^{-1/2} / \frac{1}{2} y^{-1/2} = \sqrt{\frac{y}{x}} \quad \frac{dy}{dx} = \sqrt{\frac{y}{x}}$$

(ii) $2x^2 + 3xy = 5$

differentiating with respect to x

$$4x + 3y + 3x \frac{dy}{dx} = 0 \quad \frac{dy}{dx} = \frac{-4x - 3y}{3x} = -\frac{4x + 3y}{3x}$$

(iii) $x^2 + y^2 = 3^2$ differentiating with respect to x

$$2x + 2y \frac{dy}{dx} = 0 \quad \frac{dy}{dx} = -\frac{x}{y}$$

(iv) $5y^2 = 4x$ differentiating with respect to x

$$10y \frac{dy}{dx} = 4 \quad \frac{dy}{dx} = \frac{2}{5y}$$

(v) $2x + 3y = \sqrt{7}$ differentiating with respect to x

$$2 + 3 \frac{dy}{dx} = 0 \quad \frac{dy}{dx} = -\frac{2}{3}$$

(vi) $2x^2 + 3y^2 = 25$ differentiating with respect to x

148

$$4x + 6y \frac{dy}{dx} = 0, \quad \frac{dy}{dx} = -\frac{2x}{3y}$$

13.   (i) $\dfrac{dy}{dx} = \dfrac{\sqrt{y}}{x}$ $\sqrt{x} + \sqrt{y} = \sqrt{3}$

$$= \frac{\sqrt{y}}{\sqrt{x}} \qquad \text{when } x = 1, \quad \sqrt{y} = \sqrt{3} - 1$$

$$\frac{dy}{dx} = \sqrt{3} - 1$$

(ii) $\dfrac{dy}{dx} = -\dfrac{4x + 3y}{3x}$ $2x^2 + 3xy = 5$
                                    when $x = 1, 2 + 3y = 5$

$$= -\frac{4 + 3}{3} \qquad\qquad 3y = 3$$
$$\qquad\qquad\qquad\qquad\qquad y = 1$$

$$\frac{dy}{dx} = -\frac{7}{3}$$

(iii) $\dfrac{dy}{dx} = -\dfrac{x}{y}$ $x^2 + y^2 = 3^2$

                                when $x = 1, 1 + y^2 = 9$

$$\frac{dy}{dx} = -\frac{1}{\pm 2\sqrt{2}} \qquad\qquad y^2 = 8$$

$$= \mp \frac{1}{2\sqrt{2}} \qquad\qquad y = \pm 2\sqrt{2}$$

$$= \mp 0.3535$$

(iv) $\dfrac{dy}{dx} = \dfrac{2}{5y}$ $\dfrac{2}{5\left(\dfrac{2}{\sqrt{5}}\right)} = \pm\dfrac{\sqrt{5}}{5}$ $5y^2 = 4x$

                                           when $x = 1, y^2 = 4/5$

$$y = \pm\frac{2}{\sqrt{5}}$$

$$= \pm 0.447$$

(v) $\dfrac{dy}{dx} = -2/3$ independent of the value of x

(vi) $\dfrac{dy}{dx} = -\dfrac{2x}{3y}$ $2x^2 + 3y^2 = 25$ when $x = 1$

$$= -\dfrac{2}{3(\pm 2.77)}$$ $2 + 3y^2 = 25$

$$y^2 = 23/3 \qquad y = \pm 2.77$$

$$= \mp 0.241$$

14.   (i) $y = \dfrac{x - 7}{x^2 - x + 5}$ $\dfrac{dy}{dx} = \dfrac{(x^2 - x + 5) - (x - 7)(2x - 1)}{(x^2 - x + 5)^2}$

$$\dfrac{dy}{dx} = \dfrac{x^2 - x + 5 - 2x^2 + 14x + x - 7}{(x^2 - x + 5)^2} = \dfrac{-x^2 + 14x - 2)}{(x^2 - x + 5)^2}$$

(ii) $y = \dfrac{1}{1 - x} = (1 - x)^{-1}$ $\dfrac{dy}{dx} = -(1 - x)^{-2}(-1) = \dfrac{1}{(1 - x)^2}$

(iii) $y = \dfrac{\sqrt{x + 1}}{\sqrt{x}} = 1 + \dfrac{1}{\sqrt{x}} = 1 + x^{-1/2}$ $\dfrac{dy}{dx} = -\dfrac{1}{2}x^{-3/2} = -\dfrac{1}{2x^{3/2}}$

(iv) $y = \dfrac{1}{2x^2 + 3x + 4}$ $\dfrac{dy}{dx} = \dfrac{(4x + 3)}{(2x^2 + 3x + 4)^2}$

(v) $y = \dfrac{x - 1}{\sqrt{x}} = (x - 1)(x^{-1/2} = x^{1/2} - x^{-1/2}$

$$\dfrac{dy}{dx} = \dfrac{1}{2}x^{-1/2} + \dfrac{1}{2}x^{-3/2} = \dfrac{1}{2\sqrt{x}} + \dfrac{1}{2\sqrt{x^3}}$$

15.   (i) $\dfrac{d}{dx}(5x^4 + 5x - 1) = 20x^3 + 5$

(ii) $\dfrac{d}{dx}\left(\dfrac{y - 1}{y + 1}\right) = \dfrac{y + 1 - (y - 1)}{(y + 1)^2} = \dfrac{2}{(y + 1)^2}$

(iii) $\dfrac{d}{dZ}\left(\dfrac{1}{Z}+\dfrac{1}{Z^2}+\dfrac{1}{Z^3}\right)=\dfrac{d}{Z}\left(Z^{-1}+Z^{-2}+Z^{-3}\right)$

$$=-Z^{-2}-2Z^{-3}-3Z^{-4}$$

$$=-\dfrac{1}{Z^2}-\dfrac{2}{Z^3}-\dfrac{3}{Z^4}$$

(iv) $\dfrac{d}{dt}\left(\dfrac{t^4+1}{2t}\right)=\dfrac{d}{dt}\left(\dfrac{1}{2^3}+\dfrac{1}{2}t^{-1}\right)=\dfrac{3}{2}t^2-\dfrac{1}{2\,t^2}$

(v) $\dfrac{d}{du}\{(u^2+1)\,(u^3+2)\}=$

$\dfrac{d}{du}(u^5+u^3+2u^2+2)=5u^4+3u^2+4u.$

16. (i) $y=\dfrac{1}{\sqrt{x}}-\dfrac{1}{x}+\dfrac{1}{x^2}=x^{-1/2}-x^{-1}+x^{-2}$

$\dfrac{dy}{dx}=-\dfrac{1}{2}x^{-3/2}+x^{-2}-2x^{-3}=-\dfrac{1}{2x^{3/2}}+\dfrac{1}{x^2}-\dfrac{2}{x^3}$

(ii) $y=(x+1)\,(x+2)\,(x+3)=(x^2+3x+2)\,(x+3)$

$$=x^3+3x^2+2x+3x^2+9x+6$$

$y=x^3+6x^2+11x+6 \qquad \dfrac{dy}{dx}=3x^2+12x+11$

(iii) $y=\dfrac{\sqrt{x}-1}{\sqrt{x}+1}=\dfrac{x^{1/2}-1}{x^{1/2}+1}$

$\dfrac{dy}{dx}=\dfrac{1}{2}\dfrac{x^{-1/2}(x^{1/2}+1)-(x^{1/2}-1)\dfrac{1}{2}x^{-1/2}}{(x^{1/2}+1)^2}$

$=\dfrac{\dfrac{1}{2}+\dfrac{1}{2}x^{-1/2}-\dfrac{1}{2}+\dfrac{1}{2}x^{-1/2}}{(x^{1/2}+1)^2}=\dfrac{x^{-1/2}}{(x^{1/2}+1)^2}$

$\dfrac{dy}{dx}=\dfrac{1}{\sqrt{x}\left(\sqrt{x}+1\right)^2}$

(iv) $\qquad y = \left(1 - \dfrac{1}{x^2}\right)^5 = (1 - x^{-2})^5$

$$\frac{dy}{dx} = 5\,(1 - x^{-2})^4\,(2\,x^{-3}) = \frac{10\,(1 - x^{-2})^4}{x^3} = \frac{10}{x^3}\left(1 - \frac{1}{x^2}\right)^4$$

(v) $\qquad xy + x^2\,y^2 + x^3\,y^3 = 0$ differentiating with respect to x

$$y + x\frac{dy}{dx} + 2x\,y^2 + x^2\,2y\,\frac{dy}{dx} + 3\,x^2\,y^3 + 3\,x^3\,y^2\,\frac{dy}{dx} = 0$$

$$\frac{dy}{dx} = -\,\frac{y\,(1 + 2xy + 3\,x^2\,y^2)}{x\,(1 + 2\,xy + 3\,x^2\,y^2)} = -\,\frac{y}{x} \qquad\qquad \frac{dy}{dx} = -\,\frac{y}{x}$$

17. $\qquad$ If $pv = 100$ $\qquad\qquad\qquad$ when $v = 25$

$\qquad$ (i) $\qquad \dfrac{dp}{dv}\,v + p = 100$ $\qquad\qquad p25 = 100$
$\qquad\qquad\qquad\qquad\qquad\qquad\qquad\qquad\qquad p = 4$

$$\frac{dp}{dV} = \frac{100 - p}{v}$$

$\qquad\qquad$ when $v = 25$ $\quad \dfrac{dp}{dv} = \dfrac{100 - 4}{25} = 96/25$

$\qquad$ (ii) $\qquad \dfrac{dp}{dv} = \dfrac{100 - 50}{2}$ $\qquad\qquad$ when $v = 2$

$\qquad\qquad = 25.$ $\qquad\qquad\qquad\qquad\qquad p2 = 100$
$\qquad\qquad\qquad\qquad\qquad\qquad\qquad\qquad p = 50.$

SOLUTIONS 2

1.   (i) $y = 3 \sin y + \delta y = 3 \sin (x + \delta y)$ subtracting the equations

$$\delta y = 3 \sin (x + \delta x) - 3 \sin x$$

$$= 3 \sin x \cos \delta x + 3 \sin \delta x \cos x - 3 \sin x$$

as $\delta x \to 0$, $\cos \delta x \to 1$, $\sin \delta x \to \delta x$

$\delta y = 3 \sin x + 3 \delta x \cos x - 3 \sin x \; \delta y = 3 \delta x \cos x$

dividing each side by δx $\dfrac{\delta y}{\delta x} = 3 \cos x$

as $\delta x \to 0$ $\dfrac{\delta y}{\delta x} \to \dfrac{dy}{dx}$ $\dfrac{dy}{dx} = 3 \cos x.$

(ii) $y = -2 \cos x, y + \delta y = -2 \cos (x + \delta x)$ subtracting the equations

$\delta y = -2 \cos (x + \delta x) - (-2 \cos x)$
$\delta y = -2 \cos x \cos \delta x + 2 \sin x \sin \delta x + 2 \cos x$
$\delta y = 2 \sin x \; \delta x$

as $\delta x \to 0$, $\sin \delta x \to \delta x$, $\cos \delta x \to 1$ dividing each side by δx

$$\dfrac{\delta y}{\delta x} = 2 \sin x \text{ as } \delta x \to 0, \dfrac{\delta x}{\delta y} \to \dfrac{dy}{dx}$$

$$\dfrac{dy}{dx} = 2 \sin x$$

(iii) $y = \tan 2x, y + \delta y = \tan 2 (x + \delta x)$ subtracting the two equations
$\delta y = \tan 2 (x + \delta x) - \tan 2x$

$$= \dfrac{\tan 2x + \tan 2 \delta x}{1 - \tan 2x \tan 2 \delta x} - \tan 2x \quad \text{as } \delta x \to 0 \; \tan 2 \delta x \to 2 \delta x$$

$$\delta y = \dfrac{\tan 2x + 2 \delta x}{1 - \tan 2x (2 \delta x)} - \tan 2x$$

$$= \frac{\tan 2x + 2\,\delta x - \tan 2x + \tan^2 2x\,(2\,\delta x)}{1 - \tan 2x\,(2\,\delta x)} = \frac{2\,\delta x + 2\,\delta x \tan^2 2x}{1 - (2\,\delta x)\tan 2x}$$

dividing both sides by δx

$$\frac{\delta y}{\delta x} = \frac{2\,(1 + \tan^2 2x)}{1 - (2\delta x)\tan 2x} \quad \text{as } \delta x \to 0, \quad \frac{\delta y}{\delta x} \to \frac{dy}{dx}$$

$$\frac{dy}{dx} = 2\,(1 + \tan^2 2x) = 2 \sec^2 2x \qquad \frac{dy}{dx} = 2 \sec^2 2x.$$

2.   (i) $y = \sin x,\ \dfrac{dy}{dx} = \cos x$ (ii) $y = \cos x,\ \dfrac{dy}{dx} = -2 \sin x$

       (iii) $y = 3 \tan x,\ \dfrac{dy}{dx} = 3 \sec^2 x$ (iv) $y = 4 \cot x,\ \dfrac{dy}{dx} = -4 \cos$

       (v) $y = 5 \operatorname{cosec} x,\ \dfrac{dy}{dx} = -5 \cot x \operatorname{cosec} x$

       (vi) $y = 6 \sec x,\ \dfrac{dy}{dx} = 6 \tan x \sec x$

3.   (i) $y = \dfrac{1}{2} \sin 2x \qquad \dfrac{dy}{dx} = \cos 2x$

       (ii) $y = 3 \cos \dfrac{1}{3} x \qquad \dfrac{dy}{dx} = -\sin \dfrac{1}{3} x$

       (iii) $y = 4 \tan \dfrac{1}{4} x \qquad \dfrac{dy}{dx} = \sec^2 \dfrac{1}{4} x$

       (iv) $y = \dfrac{1}{5} \operatorname{cosec} 5x, \qquad \dfrac{dy}{dx} = -\cot 5x \operatorname{cosec} 5x$

       (v) $y = 7 \sec 7x, \qquad \dfrac{dy}{dx} = 49 \sec 7x \tan 7x$

       (vi) $y = 2 \cot 3x, \qquad \dfrac{dy}{dx} = -6 \operatorname{cosec}^2 3x.$

4.   (a)   (i) $\dfrac{dy}{dx} = 1$ (ii) $\dfrac{dy}{dx} = 0$ (iii) $\dfrac{dy}{dx} =$

$$\text{(iv)} \quad \frac{dy}{dx} = -\infty \qquad \text{(v)} \quad \frac{dy}{dx} = 0 \qquad \text{(vi)} \quad \frac{dy}{dx} = -\infty$$

(b)   (i) $\quad \dfrac{dy}{dx} = \cos 2\pi/4 = 0 \qquad$ (ii) $\quad \dfrac{dy}{dx} = -\sin \dfrac{\pi}{12}$

$$= -0.259$$

(iii) $\quad \dfrac{dy}{dx} = \sec^2 \pi/16 = 1.04 \qquad$ (iv) $\quad \dfrac{dy}{dx} = -\cot \dfrac{5\pi}{4} \operatorname{cosec} \dfrac{5\pi}{4}$

$$= +1 \times 1.414$$
$$= +1.414$$

(v) $\quad \dfrac{dy}{dx} = 49 \sec \dfrac{7\pi}{4} \tan \dfrac{7\pi}{4} = 49 \times 1.414 \times (-1) = -69.3$

(vi) $\quad \dfrac{dy}{dx} = -6 \operatorname{cosec}^2 3\pi/4 = -12$

(c)   (i) $\quad \dfrac{dy}{dx} = \cos 2\,(3\pi/4) = \cos 3\pi/2 = 0$

(ii) $\quad \dfrac{dy}{dx} = -\sin \dfrac{1}{3}\, 3\pi/4 = -\sin \pi/4 = -0.707$

(iii) $\quad \dfrac{dy}{dx} = \sec^2 \dfrac{1}{4}\, x = \sec^2 3\,\pi/16 = 1.4$

(iv) $\quad \dfrac{dy}{dx} = -\cot \dfrac{15\pi}{4} \operatorname{cosec} \dfrac{15\pi}{4} = -(-1)(-1.414) = -1.414$

(v) $\quad \dfrac{dy}{dx} = 49 \sec 21\,\dfrac{\pi}{4} \tan 21\,\dfrac{\pi}{4} = 49\,(-1.414)(1) = -69.3$

(vi) $\quad \dfrac{dy}{dx} = -6 \operatorname{cosec}^2 9\pi/4 = -12.$

5.   (i) $\quad y = x \sin x, \quad \dfrac{dy}{dx} = \sin x + x \cos x$

(ii) $\quad y = x^2 \sin^2 x, \quad \dfrac{dy}{dx} = 2x \sin^2 + x^2\, 2 \sin x \cos x$

155

(iii) $\quad y = \tan^2 x, \dfrac{dy}{dx} = 2 \tan x \sec^2 x$

(iv) $\quad y = 3 \sec^2 \tan x, \dfrac{dy}{dx} = 6 \sec x \tan^2 + 3 \sec^2 x \sec^2 x$

(v) $\quad y = 5 \ \text{cosec}^2 \ x$, let $u = \text{cosec} \ x \quad \dfrac{dy}{dx} = -\text{cosec} \ x \cot x$

$$y = 5 \ u^3 \dfrac{dy}{dx} = 15 \ u^2 \ \dfrac{dy}{dx} = \dfrac{dy}{dx} = 15 \ u^2 \ (-\text{cosec} \ x \cot x)$$

$$= -15 \ \text{cosec}^3 \ x \cot x$$

(vi) $\quad y = \cot^4 x$, let $u = \cot x \quad \dfrac{du}{dx} = -\text{cosec}^2 x$

$$y = u^4 \dfrac{dy}{dx} = 4 \ u^3 \ \dfrac{dy}{dx} = \dfrac{dy}{du} \dfrac{du}{dx} = 4 \ u^3 \ (-\text{cosec}^2 x)$$

$$= -4 \cot^3 x \ \text{cosec}^2 x$$

6. (i) $\quad y = 2 \sin^{3/4} t$, let $u = \sin t \quad \dfrac{du}{dt} = \cos t, y = 2 \ u^{3/4}, \dfrac{dy}{dx} = \dfrac{3}{2} \ u^{-1}$

$$\dfrac{dy}{dt} = \dfrac{dy}{du} \dfrac{du}{dt} = \dfrac{3}{2} \ (\sin t)^{-1/4} \cos t \ \dfrac{dy}{dt} = \dfrac{3}{2} \ \dfrac{\cos t}{\sin^{1/4} t}$$

(ii) $\quad y = -\tan^{5/2} 3t$, let $u = 3t \quad y = -\tan^{5/2} u$, let $v = \tan u$

$$y = -v^{5/2}, \dfrac{du}{dt} = 3, \dfrac{dv}{du} = \sec^2 u \ \dfrac{dy}{dv} = -\dfrac{5}{2} \ v^{3/2}, \dfrac{dy}{dt} = \dfrac{dy}{dv} \dfrac{dv}{du}$$

$$= -\dfrac{5}{2} \ v^{3/2} \sec^2 u \ 3 \ \dfrac{dy}{dt} = -\dfrac{15}{2} \ \tan^{3/2} 3t \sec^2 3t.$$

(iii) $\quad y = \sqrt{\text{cosec} \ t} \quad$ let $u = \text{cosec} \ t \ \dfrac{du}{dt} = -\text{cosec} \ t \cot t$

$$y = u^{1/2} \dfrac{dy}{du} = \dfrac{1}{2} \ u^{-1/2} \ \dfrac{dy}{dt} = \dfrac{dy}{du} \dfrac{du}{dt} = \dfrac{1}{2} \ u^{-1/2} \ (-\text{cosec} \ t \ \text{co}$$

$$= -\dfrac{1}{2} \ \text{cosec}^{1/2} t \cot t \qquad = -\dfrac{1}{2} \ \sqrt{\text{cosec} \ t} \ \cot t.$$

7.   (i) $y = x^2 \sqrt{\cos x} = x^2 \cos^{1/2} x$

$$\frac{dy}{dx} \quad 2x \sqrt{\cos x} + \frac{1}{2} x^2 (\cos x)^{-1/2} (-\sin x)$$

$$= 2x \sqrt{\cos x} - \frac{1}{2} x^2 \frac{\sin x}{\sqrt{\cos x}}$$

(ii) $y = x \sqrt{\sin x} = x (\sin x)^{1/2} \quad \frac{dy}{dx} = \sqrt{\sin x} + \frac{1}{2} x (\sin x)^{1/2} \cos x$

$$= \sqrt{\sin x} + \frac{1}{2} x \frac{\cos x}{\sqrt{\sin x}}$$

(iii) $y = x \sqrt{\tan x} = x (\tan x)^{1/2} \frac{dy}{dx} = \sqrt{\tan x} + \frac{1}{2} x (\tan x)^{-1/2} \sec^2 x$

$$= \sqrt{\tan x} + \frac{1}{2} x \frac{\sec^2 x}{\sqrt{\tan x}}$$

8. $xy = \tan y$ differentiating implicitly with respect to x

$$y + x \frac{dy}{dx} = \sec^2 y \frac{dy}{dx}, \quad \frac{dy}{dx} (\sec^2 y - x) = y$$

$$\frac{dy}{dx} = \frac{y}{\sec^2 y - x} = \frac{y}{1 + \tan^2 y - x} = \frac{y}{1 + x^2 y^2 - x}$$

9.   If $y = \dfrac{\sqrt{1 - x^2}}{\cos^{-1} x} = \dfrac{(1 - x^2)^{1/2}}{\cos^{-1} x}$

$$\frac{dy}{dx} = \frac{1}{2} \frac{(1 - x^2)^{-1/2} (-2x) \cos^{-1} x - (1 - x^2)^{1/2}}{(\cos^{-1} x)^2} \left(- \frac{1}{(1 - x^2)^{1/2}} \right)$$

$$\frac{dy}{dx} = - \frac{\dfrac{x \cos^{-1} x}{\sqrt{1 - x^2}} + 1}{\cos^{-1} x^2}$$

10. $y = \dfrac{\sin^{-1} x}{\sqrt{1 - x^2}} = \dfrac{\sin^{-1} x}{(1 - x^2)^{1/2}}$

$$\frac{dy}{dx} = \frac{\dfrac{1}{(1-x^2)^{1/2}}\,(1-x^2)^{1/2} - \sin^{-1}x\left[\dfrac{1}{2}\left(1-x^2\right)^{-1/2}(-2x)\right]}{1-x^2}$$

$$= \frac{\dfrac{1+x\sin^{-1}x}{\sqrt{1-x^2}}}{1-x^2}$$

11. $\quad y = \dfrac{\tan^{-1}x}{\cot^{-1}x}\qquad \dfrac{dy}{dx} = \dfrac{\dfrac{1}{1+x^2}\cot^{-1}x - \tan^{-1}x\left(-\dfrac{1}{1+x^2}\right)}{(\cot^{-1}x)^2}$

$$= \frac{\cot^{-1}x + \tan^{-1}x}{(1+x^2)\,(\cot^{-1}x)^2}.$$

12. $\quad y = \dfrac{\operatorname{cosec}^{-1}x}{\sec^{-1}x}$

$$\frac{dy}{dx} = \frac{-\dfrac{1}{x\,(x^2-1)^{1/2}}\sec^{-1}x - \operatorname{cosec}^{-1}\dfrac{1}{x\,(x^2-1)^{1/2}}}{(\sec^{-1}x)^2}$$

$$= -\frac{\sec^{-1}x + \operatorname{cosec}^{-1}x}{x\,(x^2-1)^{1/2}\,(\sec^{-1}x)^2}.$$

13. $\quad y = \dfrac{\cos^{-1}x}{\sin^{-1}x}$

$$\frac{dy}{dx} = \frac{-\dfrac{1}{(1-x^2)^{1/2}}\sin^{-1}x - \cos^{-1}x\,\dfrac{1}{(1-x^2)^{1/2}}}{(\sin^{-1}x)^2}$$

$$= -\frac{\sin^{-1}x + \cos^{-1}x}{(1-x^2)^{1/2}}$$

14. (i) $\quad y = 3\,\text{arc}\sin 3x = 3\sin^{-1}3x$

$$y/3 = \sin^{-1}3x,\; 3x = \sin y/3,\; \frac{dy}{dx} = \frac{1}{3}\cos y/3$$

$$\frac{dx}{dy} = \frac{1}{9}\cos\frac{y}{3} = \sqrt{1 - \sin^2\frac{y}{3}} = \frac{1}{9}\sqrt{1 - 9x^2}$$

$$\frac{dy}{dx} = \frac{9}{\sqrt{1 - 9 x^2}}$$

(ii) $y = -\ \text{arc tan } 2x = -\tan^{-1} 2x$

$-\ y \tan^{-1} 2x,\ 2x = \tan (-y)$

$$2 \frac{dx}{dy} = -\sec^2 (-y)$$

$$\frac{dy}{dx} = -\frac{2}{\sec^2 (-y)} = -\frac{2}{1 + \tan^2 (-y)} = -\frac{2}{1 + (2x)^2}$$

$$= -\frac{2}{1 + 4 x^2}$$

(iii) $y = 5 \text{ arcos } 4x = 5 \cos^{-1} 4x$

$\dfrac{y}{5} = \cos^{-1} 4x,\ 4x = \cos \dfrac{y}{5},\ 4 \dfrac{dx}{dy} = -\dfrac{1}{5} \sin \dfrac{y}{5}\ \dfrac{dx}{dy} = -\dfrac{1}{20} \sin \dfrac{1}{5} y$

$$\frac{dy}{dx} = -\frac{20}{\sin \dfrac{1}{5} y} = -\frac{20}{\sqrt{1 - \cos^2 \dfrac{1}{5} y}} = -\frac{20}{\sqrt{1 - 16 x^2}} \qquad \frac{dy}{dx} = -\frac{20}{1 - 16 x^2}$$

15.   (i) $y = \sin x \cos^{-1} x$

$$\frac{dy}{dx} = \cos x \cos^{-1} x - \sin x \frac{1}{(1 - x^2)^{1/2}}$$

(ii) $y = 2 \cos x \sin^{-1} x$

$$\frac{dy}{dx} = -\ 2 \sin x \sin^{-1} + 2 \cos x . \frac{1}{\sqrt{1 - x^2}}$$

(iii) $y = 3 \tan x \cot^{-1} x$

$$\frac{dy}{dx} = 3 \sec^2 x \cot^{-1} - 3 \tan x \frac{1}{1 + x^2}$$

SOLUTIONS 3

1.   (i) $e^x = 1 + \dfrac{x}{1!} + \dfrac{x^2}{2!} + \dfrac{x^3}{3!} + \ldots$

    (ii) $e^{-x} = 1 - \dfrac{x}{1!} + \dfrac{x^2}{2!} - \dfrac{x^3}{3!} - \ldots$

    (iii) $e^{2x} = 1 + 2\dfrac{x}{1!} + \dfrac{(2x)^2}{2!} + \dfrac{(2x)^3}{3!} + \ldots = 1 + 2\dfrac{x}{1!} + \dfrac{4x^2}{2!} + \dfrac{8x^3}{3!} +$

    (iv) $e^{-3x} = 1 - 3\dfrac{x}{1!} + \dfrac{9x^2}{2!} - \dfrac{27x^3}{3!} + \ldots$

    (i) $\dfrac{d}{dx}(e^x) = 0 + \dfrac{1}{1!} + \dfrac{2x}{2!} + \dfrac{3x^2}{3!} + \ldots$

                 $= 1 + \dfrac{x}{1!} + \dfrac{x^2}{2!} + \ldots = e^x$

    (ii) $\dfrac{d}{dx}(e^{-x}) = 0 - \dfrac{1}{1!} + \dfrac{2x}{2!} - \dfrac{3x^2}{3!} + \ldots$

                 $= -1 + \dfrac{x}{1!} - \dfrac{x^2}{2!} + \ldots$

                 $= -\left(1 - \dfrac{x}{1!} + \dfrac{x^2}{2!} - \ldots\right) = -e^{-x}$

    (iii) $\dfrac{d}{dx}\left(e^{2x}\right) = 0 + \dfrac{2}{1!} + \dfrac{8x}{2!} + \dfrac{24x^2}{3!} + \ldots$

                 $= 2\left(1 + \dfrac{2x}{1!} + \dfrac{4x^2}{2!} + \ldots\right)$

                 $= 2e^{2x}$

    (iv) $\dfrac{d}{dx}(e^{-3x}) = 0 - \dfrac{3}{1!} + \dfrac{18x}{2!} - \dfrac{81x^2}{3!} + \ldots$

                 $= -3\left(1 - \dfrac{3x}{1!} + \dfrac{9x^2}{2!} - \ldots\right)$

$$= -3 e^{-3x}$$

2.   (i) $y = 3 e^x$ (ii) $y = e^{-3x}$ (iii) $y = e^{x^2}$

$$\frac{dy}{dx} = 3 e^x \qquad\qquad \frac{dy}{dx} = -3 e^{-3x} \qquad\qquad \frac{dy}{dx} = 2x\, e^{x^2}$$

      (iv) $y = e^{-3x^2}$ (v) $y = n e^{ax}$ (vi) $y = e^{\frac{1}{2} x}$

$$\frac{dy}{dx} = -6x\, e^{-3x^2} \qquad\qquad \frac{dy}{dx} = an\, e^{ax} \qquad\qquad \frac{dy}{dx} = \frac{1}{2} e^{\frac{1}{2} x}$$

      (vii) $y = \dfrac{1}{2} e^{-\frac{1}{2} x}$ $\dfrac{dy}{dx} = -\dfrac{1}{4} e^{-\frac{1}{2} x}$

3.   (i) $y = 3 x^2 e^{x^2}$ (ii) $y = e^x \sin x$ $\dfrac{dy}{dx} = e^x \sin x + e^x \cos x$

$$\frac{dy}{dx} = 6x\, e^{x^2} + 3x^2 (2x\, e^{x^2}) = 6x\, e^{x^2} + 6 x^3\, e^{x^2}$$

      (iii) $y = e^{-3x} \cos 3x$ $\dfrac{dy}{dx} = 3 e^{-3x} \cos 3x - 3 e^{-3x} \sin 3x$

      (iv) $y = 3 e^{-x} \sec x$ $\dfrac{dy}{dx} = -3 e^{-x} \sec x + 3 e^{-x} \sec x \tan x$

      (v) $y = e^{3x} (x^3 + 3)$ $\dfrac{dy}{dx} = 3 e^{3x} (x^3 + 3) + e^{3x} (3 x^2)$

4.   (i) $y = \dfrac{e^x}{\sin x}$ $\dfrac{dy}{dx} = \dfrac{e^x \sin x - e^x \cos x}{\cot x}$

      (ii) $y = \dfrac{\tan x}{e^{2x}}$ $\dfrac{dy}{dx} = \dfrac{\sec^2 x\, e^{2x} - 2 \tan x\, e^{2x}}{e^{4x}}$

      (iii) $y = \dfrac{e^{3x} \sin x}{\cot x}$

$$\frac{dy}{dx} = \frac{(3 e^{3x} \sin x + e^{3x} \cos x) \cot x - e^{3x} \sin x\, (-\cosec^2 x)}{\cot^2 x}$$

$$= \frac{3\, e^{3x} \sin x \cot x + e^{3x} \cos x \cot x + e^{3x} \sin x \, \mathrm{cosec}^2 x}{\cot^2 x}$$

5.   (i) $y = x^3\, e^{3x}$ (vii) $y = e^{x^2} \sin x^2$

$$\frac{dy}{dx} = 3\, x^2\, e^{3x} + 3\, x^3\, e^{3x} \qquad \frac{dy}{dx} = 2\, e^{x^2} \sin x^2$$

$$+ e^{x^2}\, 2x \cos x^2$$

(ii) $y = e^{-2x}\,(x^2 - 1)$ $\dfrac{dy}{dx} = -2\, e^{-2x}\,(x^2 - 1) + 2\, e^{-2x}$

(iii) $y = 3\, e^{2x} \cos 2x$ $\dfrac{dy}{dx} = 6\, e^{2x} \cos 2x - 6\, e^{2x} \sin 2x$

(iv) $y = \dfrac{e^x}{\tan x}$ $\dfrac{dy}{dx} = \dfrac{e^x \tan x - e^x \sec^2 x}{\tan^2 x}$

(viii) $y = 3\,(e^{3x})^5$ $\dfrac{dy}{dx} = 15\,(e^{3x})^4\, 3\, e^{3x} = 45\,(e^{3x})^5$

(ix) $y = e^x - e^{-x}$ $\dfrac{dy}{dx} = e^x + e^{-x}$

(x) $y = e^{2x} + e^{-2x}$ $\dfrac{dy}{dx} = 2\, e^{2x} - 2\, e^{-2x}$

(v) $y = \sin(e^{3x})$ $\dfrac{dy}{dx} = 3\, e^{3x} \cos(e^{3x})$ $\dfrac{dy}{dx} = 6\, e^{x^2}$

(xii) $y = e^{e^{-x}}$ let $u = e^{-x}$ $y = e^u$ $\dfrac{dy}{dx} = e^u \dfrac{du}{dx}$ $\dfrac{dy}{dx} = - e^{-x}\, e^{e^{-x}}$

(xiii) $y = (e^{2x} + e^{-2x})^{-3}$ $\dfrac{dy}{dx} = -3\,(e^{2x} + e^{-2x})^{-4}\,(2\, e^{2x} - 2\, e^{-2x})$

$$= - \frac{6\, e^{2x} - e^{-2x})}{e^{2x} + e^{-2x})^4}$$

(xiv) $y = 3\, a^x$ $\dfrac{dy}{dx} = 3\,(\log_e a)\, a^x$

(xv) $y = 2\ (3^x)$ $\dfrac{dy}{dx} = 2\ (\log_e 3)\ 3^x.$

(xi) $y = \dfrac{e^x + e^{-x}}{e^x - e^{-x}}$ $\dfrac{dy}{dx}\ \dfrac{(e^x - e^{-x})\ (e^x - e^{-x}) - (e^x + e^{-x})\ (e^x + e^{-x})}{(e^x - e^{-x})^2}$

$= \dfrac{e^{2x} - 2 + e^{-2x} - e^{2x} - 2 - e^{-2x}}{(e^x - e^{-x})^2} = -\dfrac{4}{(e^x - e^{-x})^2}$

6.   (i) $y = e^{nx} \sin kx$

$\dfrac{dy}{dx} = n\ e^{nx} \sin kx + e^{nx}\ k \cos kx$

(ii) $y = e^{-mx} \cos nx$ $\dfrac{dy}{dx} = -m\ e^{-mx} \cos nx - n\ e^{-mx} \sin nx$

(iii) $y = \dfrac{e^{ax}}{\cos bx}$ $\dfrac{dy}{dx} = \dfrac{a\ e^{ax} \cos bx + b\ e^{ax} \sin bx}{\cos^2 bx}$

(iv) $y = e^{2x} \cos 5x$ $\dfrac{dy}{dx} = 2\ e^{2x} \cos 5x - 5\ e^{2x} \sin 5x$

(v) $y = e^x \sin 5x$ $\dfrac{dy}{dx} = e^x \sin 5x + 5\ e^x \cos 5x.$

7.   (i) $y = e^{-1/t}$ let $u = -1/t,$ $\dfrac{dy}{dt} = 1/t^2$ $y = e^u$

$\dfrac{dy}{du} = e^u,\ \dfrac{dy}{dt} = \dfrac{dy}{du}\cdot\dfrac{du}{dt} = e^u\ \dfrac{1}{t^2}$

$\dfrac{dy}{dt} = \dfrac{1}{t^2}\ e^{-1/t}$

(ii) $y = e^{\sqrt{u}}$

let $w = \sqrt{u} = u^{1/2},\ \dfrac{dw}{du} = \dfrac{1}{2}\ u^{-1/2}$

$y = e^w$ $\dfrac{dy}{dw} = e^w,\ \dfrac{dy}{du} = \dfrac{dy}{dw}\cdot\dfrac{dw}{du} = \dfrac{1}{2}\ e^w\ u^{-1/2}$

$$= \frac{1}{2} e^{\sqrt{u}} \, u^{-1/2} = \frac{1}{2\sqrt{u}} e^{\sqrt{u}}.$$

(iii) $y = e^{-\sin x}$ let $u = -\sin x$, $\dfrac{du}{dx} = -\cos x$

$$y = e^u, \ \frac{dy}{du} = e^u, \ \frac{dy}{dx} = \frac{dy}{du}\frac{du}{dx} = -e^u \cos x \qquad \frac{dy}{dx} = -\cos x \, e^{-\sin x}.$$

$$y = e^{-x}$$

8.   (i) $y + \delta y = e^{(-x + \delta x)}$ ...(1)

$$= e^{-x} e^{-\delta x} \text{ ...(2)}$$

substracting (1) from (2)

$$\delta y = e^{-x} e^{-\delta x} - e^{-x} = e^{-x} (e^{-\delta x} - 1)$$

$$= e^{-x} \left[1 - \frac{\delta x}{1} + \frac{(\delta x)^2}{2!} - \frac{(\delta x)^3}{3!} + ... - 1 \right]$$

$$= e^{-x} \left[-\delta x + \frac{(\delta x)^2}{2!} - \frac{(\delta x)^3}{3!} + ... \right]$$

$$\delta y = \delta x \, e^{-x} \left[-1 + \frac{\delta x}{2!} - \frac{(\delta x)^2}{3!} + ... \right]$$

$$\frac{\delta y}{\delta x} = e^{-x} \left[-1 + \frac{\delta x}{2!} - \frac{(\delta x)^2}{3!} + ... \right]$$

as $\delta x \to 0$, $\dfrac{\delta y}{\delta x} \to \dfrac{dy}{dx}$

$$\frac{dy}{dx} = -e^{-x}$$

(ii) $y = e^{2x}$ ...(1) $y + \delta y = e^{2(x + \delta x)}$ ...(2)

subtracting (1) from (2)

$$\delta y = e^{2(x + \delta x)} - e^{2x} = e^{2x} e^{2\delta x} - e^{2x} = e^{2x} (e^{2\delta x} - 1)$$

$$= e^{2x} \left[1 + \frac{2\delta x}{1!} + \frac{(2\delta x)^2}{2!} + \frac{(2\delta x)^3}{3!} + ... - 1 \right]$$

164

$$= e^{2x}\left(2\ \delta x + \frac{4\ \delta x^2}{2!} + \frac{8\ \delta x^3}{3!} + \ldots\right)$$

$$\delta y = \delta x\ e^{2x}\left(2 + \frac{2\ \delta x}{2!} + \frac{4\ \delta x^2}{3!} + \ldots\right)$$

dividing each side by δx

$$\frac{\delta y}{\delta x} = e^{2x}\left(2 + \frac{2\ \delta x}{2!} + \frac{4\ \delta x^2}{3!} + \ldots\right) \text{ as } \delta x \to 0,\ \frac{\delta y}{\delta x} \to \frac{dy}{dx}$$

$$\frac{dy}{dx} = 2\ e^{2x}.$$

9.   (i) $y = 2\ e^x$ (ii) $y = 3\ e^{-x}$

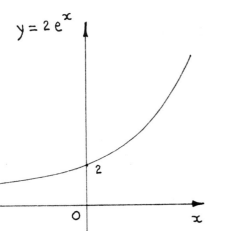

Fig. 28

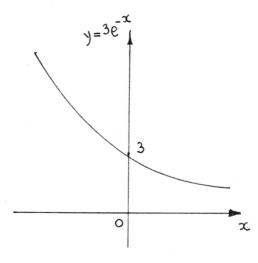

Fig. 29

10.　　$y = e^{1/x}$ ...(1)

$y + \delta y = e^{1/(x + \delta x)}$ ...(2)

subtracting (1) from (2)

$$\delta y = e^{\frac{1}{x + \delta x}} - e^{\frac{1}{x}} \quad = 1 + \frac{1}{x + \delta x} + \frac{1}{(x + \delta x)^2 2!} + ... - 1 - \frac{1}{x} - \frac{1}{x^2 2!} -$$

$$= \frac{1}{x + \delta x} - \frac{1}{x} + \frac{1}{(x + \delta x)^2} - \frac{1}{x^2 2!} + ...$$

$$= \frac{x + x - \delta x}{x (x + \delta x)} + \frac{x^2 - x^2 - 2x \delta x - \delta x^2}{x^2 (x + \delta x)^2 2!} \quad \delta y = \frac{- \delta x}{x (x + \delta x)} + \frac{2x \delta x - \delta^2}{x^2 (x + \delta x)^2 2}$$

dividing by δx each side

$$\frac{\delta y}{\delta x} = - \frac{1}{x (x + \delta x)} - \frac{2x - \delta x}{x^2 (x + \delta x)^2 2!} + ...$$

$$\frac{dy}{dx} = - \frac{1}{x^2} - \frac{2x}{x^4 2!} - ... = - \frac{1}{x^2} \left(1 + \frac{1}{x} + \frac{1}{x^2} 2! + ... \right) = - \frac{1}{x^2} e^{1/x}.$$

as $\delta x \to 0 \quad \dfrac{\delta y}{\delta x} \to \dfrac{dy}{dx}$

SOLUTIONS 4

1.   (i) $y = \log |x|$ $y = \dfrac{\log_e |x|}{\log_e 10}$

                   $\dfrac{dy}{dx} = \dfrac{1}{x \log_e 10}$

2.   (ii) $y = 3 \ln |x|$ $\dfrac{dy}{dx} = \dfrac{3}{x}$

         (iii) $y = (3x)^x$ $\log_e y \log_e (3x)^x = \log_e 3x$

                   $\dfrac{1}{y} \dfrac{dy}{dx} = \log_e 3x + x \dfrac{1}{x}$ $\dfrac{dy}{dx} = y (\ln 3x + 1) = (3x)^x (\ln 3x + 1)$

         (iv) $y = 7^x$ $\ln y = x \ln 7$ $\dfrac{1}{y} \dfrac{dy}{dx} = \ln 7$ $\dfrac{dy}{dx} = y \ln 7 = 7^x \ln 7$

         (v) $y = \ln |kx|$ $\dfrac{dy}{dx} = \dfrac{1}{x}$

2.   (i) $y = (\cos x)^x$ $\ln y = x \ln \cos x$

                   $\dfrac{1}{y} \dfrac{dy}{dx} = \ln \cos x + x \dfrac{1}{\cos x} (- \sin x)$

                   $\dfrac{dy}{dx} = (\cos x)^x \left(\ln \cos x - \dfrac{\sin x}{\cos x} \right)$

         (ii) $y = (\cot x)^x$ $\ln y = x \ln \cot x$

                   $\dfrac{1}{y} \dfrac{dy}{dx} = \ln \cot x + x \dfrac{1}{\cot x} (- \operatorname{cosec}^2 x)$

                   $\dfrac{dy}{dx} = (\cot x)^x \left(\ln \cot x - x \dfrac{\operatorname{cosec}^2 x}{\cot x} \right)$

         (iii) $y = (x + 1)^x$ $\ln y = x \ln (x + 1)$

167

$$\frac{1}{y}\frac{dy}{dx} = \ln(x+1) + \frac{x}{1+x}$$

$$\frac{dy}{dx} = (x+1)^x \left[\ln(x+1) + \frac{x}{1+x}\right]$$

(iv) $y = \sqrt{\dfrac{x^2+1}{(x^3-1)(x^4+1)}} = (x^2+1)^{1/2}(x^3-1)^{-1/2}(x^4+1)^{-1/2}$

$$\ln y = \frac{1}{2}\ln(x^2+1) - \frac{1}{2}\ln(x^3-1) - \frac{1}{2}\ln(x^4+1)$$

$$\frac{1}{y}\frac{dy}{dx} = \frac{2x}{2(x^2+1)} - \frac{3x^2}{2(x^3-1)} - \frac{4x^3}{2(x^4+1)}$$

$$\frac{dy}{dx} = (x^2+1)^{1/2}(x^3-1)^{-1/2}(x^4+1)^{-1/2}$$

$$\left[\frac{x}{x^2+1} - \frac{3}{2}\frac{x^2}{(x^3-1)} - \frac{2x^3}{x^4+1}\right]$$

(v) $y = \sqrt[3]{\dfrac{(x-1)}{(x+1)(x+\backslash 2)}} = (x-1)^{1/3}(x+1)^{-1/3}(x+2)^{-1/3}$

$$\ln y = \frac{1}{3}\ln(x-1) - \frac{1}{3}\ln(x+1) - \frac{1}{3}\ln(x+2)$$

$$\frac{1}{y}\frac{dy}{dx} = \frac{1}{3}(x-1) - \frac{1}{3}(x+1) - \frac{1}{3}(x+2)$$

$$\frac{dy}{dx} = (x-1)^{1/3}(x+1)^{-1/3}(x+2)^{-1/3}\left[\frac{1}{3}(x-1) - \frac{1}{3}(x-1) - \frac{1}{3}(x+\right.$$

3.   (i) $y = \ln 5\,x^{1/5} = \ln 5 + \dfrac{1}{5}\ln x \qquad \dfrac{dy}{dx} = \dfrac{1}{5x}$

(ii) $y = \ln\left|\dfrac{1-x}{x}\right| = \ln|1-x| - \ln|x| \qquad \dfrac{dy}{dx} = \dfrac{1}{1-x} - \dfrac{1}{x}$

(iii) $y = x^2 \ln x \qquad \dfrac{dy}{dx} = 2x\ln x + x^2\dfrac{1}{x} = 2x\ln x + x$

4. (i) $y = x^{-3} \ln 3x$ $\dfrac{dy}{dx} = -3x^{-4} \ln 3x + x^{-3} \dfrac{1}{x}$

$$= -\dfrac{3}{x^4} \ln 3x + \dfrac{1}{x^4}$$

(ii) $y = x - \ln x$ $\dfrac{dy}{dx} = 1 - \dfrac{1}{x}$

(iii) $y = \dfrac{x}{\ln x}$ $\dfrac{dy}{dx} = \ln \dfrac{x - x\dfrac{1}{x}}{(\ln x)^2} = \dfrac{\ln x - 1}{(\ln x)^2}$

(iv) $y = \dfrac{\ln 2x}{\sin 2x}$ $\dfrac{dy}{dx} = \dfrac{1}{x} \sin 2x - \ln 2x (2 \cos 2x)$

(v) $y = \dfrac{\tan x}{\ln \left| \dfrac{1}{x} \right|}$

$$\dfrac{dy}{dx} = \dfrac{\sec^2 x \ln \left| \dfrac{1}{x} \right| - x \tan x}{\left[\ln \left| \dfrac{1}{x} \right| \right]^2}$$

5. $y = e^x \ln x$ $\dfrac{dy}{dx} = e^x \ln x + \dfrac{e^x}{x}$

6. $y = e^{\sin x} \cos (\ln x)$

$$\dfrac{dy}{dx} = \cos x \, e^{\sin x} \cos (\ln x) + e^{\sin x} (- \sin (\ln x)) \dfrac{1}{x}$$

7. $y = e^{\cos x} \ln (\sin x)$

$$\dfrac{dy}{dx} = - \sin x \, e^{\cos x} \ln (\sin x) + e^{\cos x} \dfrac{1}{\sin x} \cos x$$

8. $y = \log_e \sec^2 (5x - 1)$

$$\dfrac{dy}{dx} = \dfrac{5 \times 2 \sec^2 (5x - 1) \tan (5x - 1)}{\sec^2 (5x - 1)} \qquad 10 \tan (5x - 1$$

since $\quad \sec(5x - 1) = u, \dfrac{du}{dx} = 5 \sec(5x - 1) \tan(5x - 1)$

$y = \ln u^2 = 2 \ln u$

$$\dfrac{dy}{du} = \dfrac{2}{u}, \dfrac{dy}{dx} = \dfrac{dy}{du}\dfrac{du}{dx} = \dfrac{10 \sec(5x - 1) \tan(5x - 1)}{\sec(5x - 1)}$$

$$= 10 \tan(5x - 1)$$

9. (i) $\quad y = \sqrt{x(x - 1)(x + 2)}$

Taking logarithms to the base e on both sides.

$$\ln y = \dfrac{1}{2} \ln x + \dfrac{1}{2} \ln(x - 1) + \dfrac{1}{2} \ln(x + 2)$$

differentiating with respect to x.

$$\dfrac{1}{y}\dfrac{dy}{dx} = \dfrac{1}{2x} + \dfrac{1}{2}(x - 1) + \dfrac{1}{2}(x + 2)$$

$$\dfrac{dy}{dx} = \sqrt{x(x - 1)(x + 2)} \left[\dfrac{1}{2}x + \dfrac{1}{2}(x - 1) + \dfrac{1}{2}(x + 2)\right].$$

(ii) $\quad y = \sqrt{(x + 1)(x + 3)(x + 4)}$

$$\ln y = \dfrac{1}{2} \ln(x + 1) + \dfrac{1}{2} \ln(x + 3) + \dfrac{1}{2} \ln(x + 4)$$

$$\dfrac{1}{y}\dfrac{dy}{dx} = \dfrac{1}{2}(x + 1) + \dfrac{1}{2}(x + 3) + \dfrac{1}{2}(x + 4)$$

$$\dfrac{dy}{dx} = \sqrt{(x + 1)(x + 3)(x + 4)} \left[\dfrac{1}{2}(x + 1) + \dfrac{1}{2}(x + 3) + \dfrac{1}{2}(x + 4\right.$$

10. $$y = \sqrt{\frac{3\,x^3 - 4}{5\,x^3 + 7}}$$

$$\ln y = \frac{1}{2}\ln(3\,x^3 - 4) - \frac{1}{2}\ln(5\,x^3 + 7)$$

$$\frac{1}{y}\frac{dy}{dx} = \frac{1}{2} \times \frac{9\,x^2}{3\,x^3 - 4} - \frac{1}{2} \times \frac{15\,x^2}{5\,x^3 + 7}$$

$$\frac{dy}{dx} = \sqrt{\frac{3\,x^3 - 4}{5\,x^3 + 7}} \quad \left[\frac{1}{2}\left(\frac{9\,x^2}{3\,x^3 - 4} - \frac{15\,x^2}{5\,x^3 + 7}\right)\right].$$

SOLUTIONS 5

1.   (i) $y = \sin hx$ ...(1)

              $y + \delta y = \sinh (x + \delta x)$ ...(2)

subtracting (1) from (2)

$$\delta y = \sinh (x + \delta x) - \sinh x$$
$$= \sinh x \cosh \delta x + \sinh \delta x \cosh x - \sinh x$$

$$\cosh \delta x = 1 + \frac{\delta x^2}{2!} + \frac{\delta x^4}{4!} + \ldots$$

$$\sinh \delta x = \delta x + \frac{\delta x^3}{3!} + \frac{\delta x^5}{5!} + \ldots$$

$$\delta y = \sinh x \left(1 + \frac{\delta x^2}{2!} + \frac{\delta x^4}{4!} + \ldots\right) + \left(\delta x + \frac{\delta x^3}{3!} + \frac{\delta x^5}{5!} + \ldots\right) \text{co}$$

$$- \sinh x$$

as $\delta x \to 0$

$$\delta y = \sinh x + \left(\delta x + \frac{\delta x^3}{3!} + \frac{\delta x^5}{5!} + \ldots\right) \cosh x - \sinh x$$

$$= \left(\delta x + \frac{\delta x^3}{3!} + \frac{\delta x^5}{5!} + \ldots\right) \cosh x$$

$$\frac{\delta y}{\delta x} = \left(1 + \frac{\delta x^2}{3!} + \frac{\delta x^4}{5!} + \ldots\right) \cosh x$$

$$\delta x \to 0 \quad \frac{dy}{dx} = \cosh x$$

    (ii) $y = \cosh x \, \dfrac{e^x + e^{-x}}{2}$

$$y = \frac{e^x + e^{-x}}{2}$$

172

$$y + \delta y = \frac{e^{x + \delta x} + e^{-(x + \delta x)}}{2} - \frac{e^x}{2} - \frac{e^{-x}}{2}$$

$$e^{\delta x} = 1 + \frac{\delta x}{1!} + \frac{\delta x^2}{2!} + \ldots$$

$$\delta y = \frac{1}{2} e^x \left(1 + \frac{\delta x}{1!} + \frac{\delta x^2}{2!} + \ldots \right) + \frac{1}{2} e^{-x} \left(1 - \frac{\delta x}{1!} + \frac{\delta x^2}{2!} - \ldots \right)$$

$$- \frac{e^x}{2} - \frac{e^{-x}}{2}$$

$$= \frac{1}{2} e^x \left(\frac{\delta x}{1!} + \frac{\delta x}{2!} + \ldots \right) + \frac{1}{2} e^{-x} \left(- \frac{\delta x}{1!} + \frac{\delta x^2}{2!} - \ldots \right)$$

$$\frac{\delta y}{\delta x} = \frac{1}{2} e^x \left(\frac{1}{1!} + \frac{\delta x}{2!} + \ldots \right) + \frac{1}{2} e^{-x} \left(- \frac{1}{1!} + \frac{\delta x}{2!} - \ldots \right)$$

as $\delta x \to 0 \quad \dfrac{\delta y}{\delta x} \to \dfrac{dy}{dx}$

$$\frac{dy}{dx} = \frac{1}{2} e^x - \frac{1}{2} e^{-x} = \sinh x$$

$$\frac{dy}{dx} = \sinh x$$

(iii) $\qquad y = \sinh \dfrac{1}{2} x$

$$y + \delta y = \sinh \frac{1}{2} (x + \delta x) = \frac{e^{(x + \delta x)/2} - e^{(x + \delta x/2}}{2}$$

$$\delta y = \frac{1}{2} e^{\frac{1}{2}x} . e^{\frac{1}{2}\delta x} - \frac{1}{2} e^{-\frac{1}{2}x} . e^{-\frac{1}{2}\delta x} - \sinh \frac{1}{2} x$$

$$= \frac{1}{2} e^{\frac{1}{2}x} \left(1 + \frac{1}{2} \delta x + \frac{\left(\frac{1}{2} \delta x \right)^2}{2!} + \frac{\left(\frac{1}{2} \delta x \right)^3}{3!} + \ldots \right) - \frac{1}{2} e^{-\frac{1}{2}x}$$

$$\left(1 - \frac{1}{2} \delta x + \frac{\left(\frac{1}{2} \delta x \right)^2}{2!} - \ldots \right)$$

$$- \frac{1}{2} e^{\frac{1}{2}x} + \frac{1}{2} e^{-\frac{1}{2}x}$$

$$= \frac{1}{2} e^{\frac{1}{2}x} \left(\frac{1}{2} \delta x + \frac{\left(\frac{1}{2}\delta x\right)^2}{2!} + \frac{\left(\frac{1}{2}\delta x\right)^3}{3!} + \dots \right) +$$

$$\frac{\delta y}{\delta x} = \frac{1}{2} e^{\frac{1}{2}x} \left(\frac{1}{2} + \frac{1}{8} \delta x + \frac{1}{48} \delta x^2 \right) + \frac{1}{2 \times 2} e^{-\frac{1}{2}x} - \frac{1}{8 \times 2} \delta x + \dots$$

as $\delta x \to 0$ $\quad \frac{\delta y}{\delta x} \to \frac{dy}{dx}$ $\qquad\qquad$ $\frac{dy}{dx} = \frac{1}{2} \frac{1}{2} e^{1/2x} + \frac{1}{2 \times 2} e^{-\frac{1}{2}x}$

$$= \frac{1}{2} \left(\frac{1}{2} e^{\frac{1}{2}x} + \frac{1}{2} e^{-\frac{1}{2}x} \right)$$

$$= \frac{1}{2} \cosh \frac{1}{2} x$$

2. $\qquad y = (\sinh^{-1} 3x)^2$

$\qquad y = u^2$ $\qquad\qquad\qquad\qquad u = \sinh^{-1} 3x$

$$\frac{dy}{du} = 2u \qquad\qquad\qquad \frac{du}{dx} = \frac{3}{(1 + 9x^2)^{1/2}}$$

$$\frac{dy}{dx} = \frac{dy}{du} \cdot \frac{du}{dx} = \frac{2u \cdot 3}{(1 + 9 x^2)^{1/2}} = \frac{6 \sinh^{-1} 3x}{(1 + 9 x^2)^{1/2}}$$

$$\left(\frac{dy}{dx} \right)^2 = \frac{36 \, (\sin^{-1} 3x)^2}{1 + 9 x^2}$$

$$(1 + 9 x^2) \left(\frac{dy}{dx} \right)^2 = 36 \, (\sinh^{-1} 3x)^2 = 36 \, y.$$

3. $\qquad$ (i) $\qquad y = \tan 2x \coth 3x$

$$\frac{dy}{dx} = 2 \sec^2 2x \coth 3x - 3 \tan 2x \operatorname{cosech}^2 3x$$

$\qquad$ (ii) $\qquad y = \sinh 3x \cot 2x$

$$\frac{dy}{dx} = 3 \cosh 3x \cot 2x - 2 \sinh 3x \operatorname{cosec}^2 2x$$

(iii) $y = \operatorname{cosech} \dfrac{1}{x}$

$$\frac{dy}{dx} = -\frac{1}{x^2}\left(-\coth\frac{1}{x}\operatorname{cosech}\frac{1}{x}\right) = +\frac{1}{x^2}\coth\frac{1}{x}\operatorname{cosech}\frac{1}{x}$$

(iv) $y = \operatorname{sech} x^2 \qquad \dfrac{dy}{dx} = 2x\,(-\tanh x^2\,\operatorname{sech} x^2)$

$$= -2x\tanh x^2\,\operatorname{sech} x^2$$

(v) $y = 3\sinh^5 x/2 \quad$ let $u = x/2 \qquad \dfrac{dy}{dx} = \dfrac{1}{2}$

$y = 3\sinh^5 u \quad$ let $\sinh u = w \qquad \dfrac{dw}{du} = \cosh u$

$y = 3 w^5 \qquad \dfrac{dy}{dw} = 15 w^4$

$$\frac{dy}{dx} = \frac{dy}{dw}\frac{dw}{du}\frac{du}{dx} = 15 w^4\cosh u\left(\frac{1}{2}\right)$$

$$= \frac{15}{2}\sinh^4\frac{x}{2}\cosh\frac{x}{2}.$$

(vi) $y = \coth^{1/2} x\,\sinh^{3/2} x$

$$\frac{dy}{dx} = \frac{1}{2}\coth^{-1/2} x\,(-\operatorname{cosech}^2 x)\,\sinh^{3/2} x$$

$$+ \coth^{1/2} x\left(\frac{3}{2}\sinh^{1/2} x\right)\cosh x$$

$$= -\frac{1}{2}\frac{\operatorname{cosech}^2 x}{\coth^{1/2} x}\sinh^{3/2} x + \frac{3}{2}\cosh^{3/2} x\,\sinh^{3/2} x$$

4. (i) $y = 3\sinh^{-1}\dfrac{1}{x}$

$$\frac{dy}{dx} = -\frac{3}{x^2}\frac{1}{(1+(1/x)^2)^{1/2}} = -\frac{3}{(1+x^2)^{1/2}}$$

(ii) $y = \cosh^{-1} x^2$

$$\frac{dy}{dx} = 2x \frac{1}{(x^4 - 1)^{1/2}}$$

(iii) $y = 5 \cosh^{-1} (x^2 - 3x + 2)$

$$\frac{dy}{dx} = 5 (2x - 3) \frac{1}{\left[(x^2 - 3x + 2)^2 - 1\right]^{1/2}}$$

(iv) $y = \text{cosech}^2 \frac{x}{2} \, \text{sech}^2 \frac{x}{3}$

$$\frac{dy}{dx} = \frac{1}{2} 2 \, \text{cosech} \frac{x}{2} \left(- \coth \frac{x}{2} \, \text{cosech} \frac{x}{2} \right) \text{sech}^2 \frac{x}{3}$$

$$+ \text{cosech}^2 \frac{x}{2} \frac{2}{3} \, \text{sech} \frac{x}{3} \left(- \tanh \frac{x}{3} \, \text{sech} \frac{x}{3} \right)$$

$$\frac{dy}{dx} = - \coth \frac{x}{2} \, \text{cosech}^2 \frac{x}{2} \, \text{sech}^2 \frac{x}{3} - \frac{2}{3} \, \text{cosech}^2 \frac{x}{2} \, \text{sech}^2 \frac{x}{3} \tan$$

5. $y = \text{sech}^{-1} 2x$

$$\frac{dy}{dx} = - 2 \frac{1}{2x (1 - 4 x^2)^{1/2}} = - \frac{1}{x (1 - 4x^2)^{1/2}} = - x^{-1} (1 - 4 x^2)^{-1/2}$$

$$\frac{d^2 y}{dx^2} = x^{-2} (1 - 4 x^2)^{1/2} - x^{-1} \left(- \frac{1}{2} \right) (1 - 4 x^2)^{-3/2} (- 8x)$$

$$\frac{1}{x^2 (1 - 4 x^2)^{1/2}} - \frac{4}{(1 - 4 x^2)^{3/2}}.$$

6. $y = \sinh^{-1} \frac{1}{2} x$ $\dfrac{dy}{dx} = \dfrac{1}{2} \dfrac{1}{\left(1 + \frac{1}{4} x^2 \right)^{1/2}}$

$$\frac{d^2 y}{dx^2} = \frac{1 \times 2 \times \frac{1}{2} \left(1 + \frac{1}{4} x^2 \right) \left(\frac{2}{4} x \right)}{\left(1 + \frac{1}{4} x^2 \right)^2} = \frac{1}{2} \frac{x}{\left(1 + \frac{1}{4} x^2 \right)}$$

7. $\dfrac{d}{dx} (\text{cosech}^{-1} 2x) = \dfrac{du}{dx}$ where $u = \text{cosech}^{-1} 2x$ hence $2x = \text{cosech } u$

$$2 \frac{dx}{du} = - \coth u \ \text{cosech} \ u$$

but $1 - \coth^2 u = - \text{cosech}^2 u \qquad 1 + \text{cosech}^2 u = \coth^2 u$

$$2 \frac{dx}{du} = - \sqrt{1 + \text{cosech}^2 u} \quad 2x = - \sqrt{1 + 4 x^2} \ 2x$$

therefore, $\quad du/dx = - \dfrac{1}{x \sqrt{1 + 4 x^2}}, \quad \dfrac{d \ (\text{cosech}^{-1} 2x)}{dx} = - \dfrac{1}{x \sqrt{1 + 9 x^2}}$

8. $\quad \dfrac{d}{dx} (\tanh^{-1} 3x) \qquad$ let $u = \tanh^{-1} 3x \quad 3x = \tanh u$

$\qquad \qquad \qquad \dfrac{du}{dx} = \dfrac{3}{\text{sech}^2 u} \ $ but $\ 1 - \tanh^2 u = \text{sech}^2 u$

$3 \dfrac{dx}{du} = \text{sech}^2 u$

$\qquad \qquad \qquad \qquad \qquad \qquad \qquad 1 - (3x)^2 = \text{sech}^2 u$

$\dfrac{du}{dx} = \dfrac{3}{1 - 9x^2} \quad \dfrac{d}{dx} (\tanh^{-1} 3x) = \dfrac{3}{1 - 9 x^2}.$

9. (i) $\quad y = \sinh x, \dfrac{dy}{dx} = \cosh x; \text{at } x = - 5, \dfrac{dy}{dx} = \cosh (- 5) = 74.21$

at $x = 0 \ \dfrac{dy}{dx} = \cosh 0 = 1, \text{ and } \ $ at $x = 5, \cosh 5 = 74.21.$

(ii) $\quad y = \cosh x, \dfrac{dy}{dx} = \sinh x; \text{at } x = 5, \dfrac{dy}{dx} = \sinh (5) = - 74.2$

at $x = 0 \ \dfrac{dy}{dx} = \sinh 0 = 0, \text{ and } \ $ at $x = 5, \dfrac{dy}{dx} = \sinh 5 = 74.2.$

(iii) $\quad y = \tanh x, \dfrac{dy}{dx} = \text{sech}^2 x; \text{at } x = - 5, \dfrac{dy}{dx} = \sec^2 (- 5) = 1.82 \times 10^{-4}$

at $x = 0 \ \dfrac{dy}{dx} = \text{sech}^2 0 = 1, \text{and} \ $ at $x = 5, \dfrac{dy}{dx} = \text{sech}^2 5 = 1.82 \times 10^{-4}.$

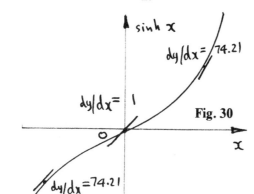

Fig. 30

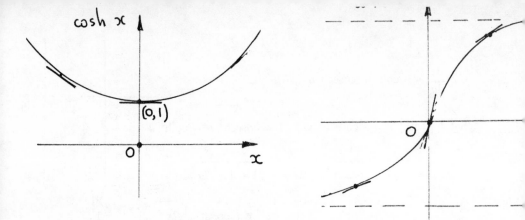

Fig. 31

Fig 32

10. $y = ar \sinh x$

$$\frac{dy}{dx} = \frac{1}{(1 + x^2)^{1/2}}, \text{ at } x = 2, \frac{dy}{dx} = \frac{1}{5^{1/2}} = 0.447$$

$$\text{at } x = 1, \frac{dy}{dx} = \frac{1}{\sqrt{2}} = 0.707$$

$y = ar \cosh x$

$$\frac{dy}{dx} = \frac{1}{(x^2 - 1)^{1/2}}, \text{ at } x = 2, \frac{dy}{dx} = \frac{1}{\sqrt{3}} = 0.577$$

$$\text{at } x = 1, \frac{dy}{dx} = \infty$$

$y = ar \tanh x$

$$\frac{dy}{dx} = \frac{1}{1 - x^2}, \text{ at } x = 2, \frac{dy}{dx} = -\frac{1}{3} = -0.333$$

$$\text{at } x = 1, \frac{dy}{dx} = \infty$$

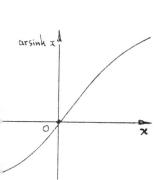

Fig. 33

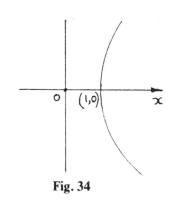

Fig. 34

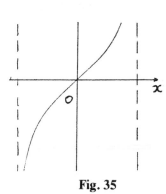

Fig. 35

11.　(i)　　$y = \sinh 2x \operatorname{cosech} 3x$

$$\frac{dy}{dx} = 2 \cosh 2x \operatorname{cosech} 3x + \sinh 2x \, (- 3 \coth 3x \operatorname{cosech} 3x)$$

$$= 2 \cosh 2x \operatorname{cosech} 3x - 3 \sinh 2x \coth 3x \operatorname{cosech} 3x$$

(ii)　　$y = \sinh 3x$ 　　　　　　$\dfrac{dy}{dx} = 3 \cosh 3x$

(iii)　　$y = e^x \cosh 2x$ 　　　　$\dfrac{dy}{dx} = e^x \cosh 2x + 2\, e^x \sinh 2x$

(iv)　　$y = \ln \sinh 5x$ 　　　　$\dfrac{dy}{dx} = \dfrac{5 \cosh 5x}{\sinh 5x} = 5 \coth 5x$

(v)　　$y = e^{\coth^2 x}$ 　　　　$\dfrac{dy}{dx} = 2 \coth x \, (- \operatorname{cosech}^2 x)\, e^{\coth^2 x}$

$$= - 2 \coth x \operatorname{cosech}^2 e^{\coth^2 x}$$

(vi)　　$y = x^3 \coth^3 5x$ 　　$\dfrac{dy}{dx} = 3\, x^2 \coth^3 5x + x^3 \, (15 \coth^2 5x)\, (- \operatorname{cosech}^2 x)$

$$= 3\, x^2 \coth^3 5x + 15\, x^3 \coth^2 5x \operatorname{cosech}^2 x$$

(vii)　　$y = \sqrt{\coth 3x} = (\coth 3x)^{1/2}$

$$\frac{dy}{dx} = \frac{1}{2}\, 3 (\coth 3x)^{-1/2} \, (- \operatorname{cosech}^2 3x) \quad = - \frac{3}{2} \frac{\operatorname{cosech}^2 3x}{(\coth 3x)^{1/2}}$$

179

(viii) $\quad y = \dfrac{1}{3} \cosh^3 x - \cosh x$

$$\dfrac{dy}{dx} = \cosh^2 x \,(\sinh x) - \sinh x$$

$$= \sinh x \,(\cosh^2 x - 1) = \sinh x \,\sinh^2 x = \sinh^3 x$$

(ix) $\quad y = 2 \tanh x \,\text{sech}^2 x$

$$\dfrac{dy}{dx} = 2 \,\text{sech}^2 \,\text{sech}^2 x + 2 \tanh x \,2 \,\text{sech}\, x \,(- \,\text{sech}\, x \tanh x)$$

$$= 2 \,\text{sech}^4 x - 4 \,\text{sech}^2 x \,\tanh^2 x$$

(x) $\quad \sqrt{\dfrac{\cosh 2x + 1}{\cosh 2x - 1}}$

$$\ln y = \dfrac{1}{2} \ln (\cosh 2x + 1) - \dfrac{1}{2} \ln (\cosh 2x - 1)$$

$$\dfrac{1}{y} \dfrac{dy}{dx} = \dfrac{1}{2} \dfrac{2 \sinh 2x}{\cosh 2x + 1} - \dfrac{1}{2} \dfrac{2 \sinh 2x}{\cosh 2x - 1}$$

$$\dfrac{dy}{dx} = \sqrt{\dfrac{\cosh 2x + 1}{\cosh 2x - 1}} \left(\dfrac{\sinh 2x}{\cosh 2x + 1} - \dfrac{\sinh 2x}{\cosh 2x - 1} \right)$$

$$= \sqrt{\dfrac{\cosh 2x + 1}{\cosh 2x - 1}} \left(\dfrac{\cosh 2x - 1 - \cosh 2x - 1}{\cosh^2 2x - 1} \right) \sinh 2x$$

$$= \sqrt{\dfrac{\cosh 2x + 1}{\cosh 2x - 1}} \left(- \dfrac{2}{\sinh 2x} \right)$$

12. (i) $\quad y = ar\sinh (\cosh 3x) = \tanh^{-1} (\cosh 3x)$

$$\dfrac{dy}{dx} = \dfrac{3 \sinh 3x}{1 - \cosh^2 3x} = - \dfrac{3 \sinh 3x}{\sinh^2 3x} = - 3 \,\text{cosech}\, 3x$$

(ii) $\quad ar\text{cosech} (\coth 2x) = \text{cosech}^{-1} (\coth 2x) = y$

180

$$\frac{dy}{dx} = \frac{\operatorname{cosech}^2 x}{\coth 2x \, (1 + \coth^2 2x)^{1/2}}$$

(iii) $y \, ar\mathrm{sech} \, (\tanh x) = \mathrm{sech}^{-1} (\tanh x)$

$$\frac{dy}{dx} = - \frac{\mathrm{sech}^2 x}{\tanh x} (1 - \tanh^2)^{1/2} = - \frac{\mathrm{sech}^2 x}{\tanh x} \, \mathrm{sech} \, x$$

$$= - \frac{\mathrm{sech} \, x}{\tanh x} = - \operatorname{cosech} x$$

$\cosh^2 - \sinh^2 x = 1 \quad 1 - \tanh^2 x = \mathrm{sech}^2 x \quad 1 - \coth^2 x = - \operatorname{cosech}^2 x$

(iv) $y = ar\tanh (\sinh x) = \tanh^{-1} \sinh x \qquad \dfrac{dy}{dx} = \dfrac{1}{1 - \sinh^2 x}$

(v) $y = ar\coth (3 x^2 - 1) = \coth^{-1} (3 x^2 - 1)$

$$\frac{dy}{dx} = \frac{1}{1 - (3 x^2 - 1)^2} \, 6x = \frac{6x}{(1 - 3 x^2 + 1)(1 + 3 x^2 - 1)}$$

$$= \frac{2}{x (2 - 3 x^2)}$$

SOLUTIONS 6

1. $x = 2 \sinh t$ $dx/dt = 2 \cosh t$

          $y = 3 \cosh t$ $dy/dt = 3 \sinh t$

$$\frac{dy/dt}{dx/dt} = \frac{dy}{dx} = \frac{3}{2} \frac{\sinh t}{\cosh t} = \frac{3}{2} \tanh t.$$

2.   (a)   Differentiate $x = t - \sin t$ and $y = 1 - \cos t$,

$$dx/dt = 1 - \cos t, \; dy/dt = \sin t \; \frac{dy/dt}{dx/dt} = \sin t$$

$$\frac{dy/dt}{dx/dt} = \frac{dy}{dx} = \frac{\sin t}{1 - \cos t} = \frac{2 \sin t/2 \; \cos t/2}{1 - (2 \cos^2 t/2 - 1)}$$

$$= \frac{2 \sin t/2 \; \cos t/2}{2 (1 - \cos^2 t/2)} = \frac{2 \sin t/2 \; \cos t/2}{2 \sin^2 t/2}$$

$$= \frac{\cos t/2}{\sin t/2} = \cot t/2$$

$$\frac{d^2 y}{dx^2} = -\frac{1}{2} \operatorname{cosec}^2 t/2 \; \frac{dt}{dx} = -\frac{1}{2} \operatorname{cosec}^2 t/2 \left(\frac{1}{1 - \cos t} \right)$$

$$= -\frac{1}{2} \operatorname{cosec}^2 t/2 \; \frac{1}{2 \sin^2 t/2} = -\frac{1}{4} \operatorname{cosec}^4 t/2.$$

(b)

t	0	$\pi/6$	$\pi/4$	$\pi/3$
$\sin t$	0	0.5	0.707	0.866
$t - \sin t$	0	0.024	0.078	0.181
$\cos t$	1	0.866	0.707	0.5
$1 - \cos t$	0	0.134	0.293	0.5
t	$7\pi/6$	$4\pi/3$	$3\pi/2$	$5\pi/3$
$\sin t$	- 0.5	- 0.866	- 1	- 0.866
$t - \sin t$	4.17	5.05	5.71	6.10
$\cos t$	- 0.866	- 0.5	0	0.5
$1 - \cos t$	1.866	1.5	1	0.5

cont....

t	$\pi/2$	$2\pi/3$	$5\pi/6$	π
$\sin t$	1	0.866	0.5	0
$t - \sin t$	0.571	1.228	2.118	3.142
$\cos t$	0	- 0.5	- 0.866	- 1
$1 - \cos t$	1	1.5	1.866	2
t	$11\pi/6$	2π		
$\sin t$	- 0.5	0		
$t - \sin t$	6.26	6.28		
$\cos t$	0.866	1		
$1 - \cos t$	0.134	0		

Plotting the points of $y = 1 - \cos t$ against the points of $x = t - \sin t$.

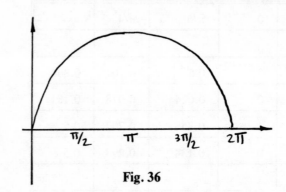

cycloid

Fig. 36

Considering $\dfrac{dy}{dx} = \cot t/2 = 0$ for turning points, $\cot t/2 = \cot \pi/2$ hence

$t = \pi$, substituting this value in the second derivative

$$\frac{d^2y}{dx^2} = -\frac{1}{4}\, \text{cosec}^4\, \frac{\pi}{2} = -\frac{1}{4} < 0$$

therefore $\dfrac{d^2y}{dx^2}$ is negative for $t = \pi$ giving a maximum.

When $t = \pi$, $y_{max} = 1 - \cos \pi = 2$.

3. $x = ct,\ dx/dt = c;\ y = c/t,\ dy/dt = -c/t^2$

(i) $\dfrac{dy}{dx} = \dfrac{dy}{dt} / \dfrac{dx}{dt} = \dfrac{-c/t^2}{c} = -\dfrac{1}{t^2}$, the gradient is always negative.

(ii) $\dfrac{d^2y}{dx^2} = t\,\dfrac{2}{t^3}\,\dfrac{dt}{dx} = \dfrac{2}{t^3}\,\dfrac{1}{c} = \dfrac{2}{c\,t^3}$

There are no turning points since $\dfrac{dy}{dx}$ cannot be zero.

Fig. 37

4. $x = 2t^2$ and $y = 2t^3$

$$\frac{dx}{dt} = 4t \qquad \frac{dy}{dt} = 6t^2 \qquad \frac{dy}{dx} = \frac{dy/dt}{dx/dt} = \frac{6t^2}{4t} = \frac{3}{2}t$$

the curve has a turning point at $t = 0$

$$\frac{d^2y}{dx^2} = \frac{3}{2}\frac{dt}{dx} = \frac{3}{2}\left(\frac{1}{4t}\right) = \frac{3}{8t}$$

$t = 0$; $x = 0$ and $y =$ 0 $t = 2$; $x = 8$ and $y =$ 16
$t = 2$; $x = 8$ and $y = -$ 16 $t = 4$; $x = 32$ and $y =$ 128
$t = 4$; $x = 32$ and $y = -$ 128 $t = 3$; $x = 18$ and $y =$ 54

Fig. 38

5. $\qquad x = 1 + t^2 \qquad y = 2t - 1$

$$\frac{dx}{dt} = 2t \qquad\qquad \frac{dy}{dt} = 2$$

$$\frac{dy}{dx} = \frac{dy/dt}{dx/dt} = 2/2t = 1/t \qquad\qquad \frac{d^2y}{dx^2} = -\frac{1}{t^2}\frac{dt}{dx} = -\frac{1}{t^2}\frac{1}{2}t = -\frac{1}{2t^3}$$

$t = $ $0; x = $ 1, $y = -1$ $t = $ $1; x = $ $2, y = $ 1
$t = -1; x = $ 2, $y = -3$ $t = $ $2; x = $ $5, y = $ 3
$t = $ $2; x = $ 5, $y = -5$ $t = $ $3; x = 10, y = -7$

Fig. 39

6. $\qquad x = 4t^2, y = 4t \; t = y/4, \; x = 4(y/4)^4 \Rightarrow x = y^2/4 \Rightarrow y^2 = 40$

$$\frac{dy}{dx}2y = 4 \Rightarrow \frac{dy}{dx} = \frac{2}{y}$$

Fig. 40

7. $x = 2 \sin t$ $\dfrac{dx}{dt} = 2 \cos t$

$y = 2 \cos^3 t$ $\dfrac{dy}{dt} = 6 \cos^2 t \, (- \sin t)$

$$\frac{dy}{dx} = -\frac{6 \sin t \cos^2 t}{2 \cos t} = -3 \sin t \cos t = -\frac{3}{2} \sin 2t$$

$$\frac{d^2y}{dx^2} = -3 \cos 2t$$

$$\frac{dy}{dx} = 0 \text{ for turning points}$$

$-\dfrac{3}{2} \sin 2t = 0 = \sin 0 = \sin \pi = \sin 2\pi$ $t = 0, t = \pi/2, t = \pi$

$$\frac{d^2y}{dx^2} = -3, \qquad \frac{d^2y}{dx^2}, = 3 \qquad \frac{d^2y}{dx^2} = -3$$

   max   min   max

when $t = 0$; $x = 0, y = 2$ $t = \pi/2$; $x = 2, y = 0$
$t = \pi$; $x = 0, y = -2$

Fig. 41

8. $x = 2 \cos t + (t + 3) \sin t$

$$\frac{dx}{dt} = -2 \sin t + \sin t + (t + 3) \cos t = -\sin t + (t + 3) \cos t$$

186

$$y = 2 \sin t - (t + 3) \cos t$$

$$\frac{dy}{dx} = 2 \cos t - \cos t + (t + 3) \sin t = \cos t + (t + 3) \sin t.$$

$$\frac{dy/dt}{dx/dt} = \frac{dy}{dx} = \frac{\cos t + (t + 3) \sin t}{- \sin t + (t + 3) \cos t}$$

9. $x = 2 \cos \Theta$, $dx/d\Theta = -2 \sin \Theta$

      $y = 4 \sin \Theta$, $dy/d\Theta = 4 \cos \Theta$

$$\frac{dy/d\Theta}{dx/dO} = \frac{dy}{dx} = -2 \cot \Theta.$$

10. $x = t + e^t$, $\dfrac{dx}{dt} = 1 + e^t$ $y = 2t - e^{2t}$, $\dfrac{dy}{dt} = 2 - 2 e^{2t}$

$$\frac{dy}{dx} = \frac{dy/dt}{dx/dt} = \frac{2 (1 - e^{2t})}{1 + e^t}$$

$$\frac{d^2y}{dx^2} = \frac{\left[-4 e^{2t} (1 + e^t) - 2 (1 - e^{2t} - 2 (1 - e^{2t}) e^t \right]}{\left(1 + e^t\right)^2} \frac{dt}{dx}$$

$$= \frac{\left(-4 e^{2t} - 4 e^{3t} - 2 e^t + 2 e^{3t} \right)}{(1 + e^t)^2} \quad \frac{1}{(1 + e^t)}$$

$$= (-2 e^{3t} - 4 e^{2t} - e^t)/(1 + e^t)^3$$

11. $x = 3t$, $dx/dt = 3$ $y = 3 \ln \sec t$, $dy/dt = \dfrac{3}{\sec t} \sec t \tan t = 3 \tan t$

     (i) $\dfrac{dy/dt}{dx/dt} = \dfrac{3 \tan t}{3} = \tan t$

     (ii) $\dfrac{d^2y}{dx^2} = \sec^2 t$ $\dfrac{dt}{dx} = \dfrac{1}{3} \sec^2 t.$

12. $x = t - \sin t$, $dx/dt = 1 - \cos t$ $y = 1 - \cos t$, $dy/dt = \sin t$

$$\frac{dy/dt}{dx/dt} = \frac{dy}{dx} = \frac{\sin t}{1 - \cos t} = \frac{2 \sin t/2 \cos t/2}{1 - 2 \cos^2 t/2 + 1}$$

$$= \frac{2 \sin t/2 \cos t/2}{2 (1 - \cos^2 t/2)} = \frac{\sin t/2 \cos t/2}{\sin^2 t/2} = \cot t/2$$

SOLUTIONS 7

1. (i) $y = 3 x^2 - 5x + 7$ (ii) $x = t - 6 t^2 + 7 t^3$

$$\frac{dy}{dx} = 6x - 5 \qquad\qquad \frac{dx}{dt} = 1 - 12t + 21 t^2$$

$$\frac{d^2y}{dx^2} = 6 \qquad\qquad \frac{d^2y}{dt^2} = -12 + 42 t$$

(iii) $u = 3 v^2 + 5 v - 1$ (iv) $w = 3 z^2 - z - 4$

$$\frac{du}{dV} = 6 v + 5 \qquad\qquad \frac{dw}{dZ} = 6z - 1$$

$$\frac{d^2u}{dV^2} = 6 \qquad\qquad \frac{d^2w}{dZ^2} = 6$$

2. (i) $y = \dfrac{3 x^2 - 1}{x + 1} \qquad \dfrac{dy}{dx} = \dfrac{6x (x + 1) - (3 x^2 - 1) (1)}{(x + 1)^2} = \dfrac{3 x^2 + 6 x + 1}{(x + 1)^2}$

$$\frac{d^2y}{dx^2} = \frac{(6x + 6) (x + 1)^2 - (3 x^2 + 6x + 1) 2 (x + 1)}{(x + 1)^4}$$

$$= \frac{(6x + 6) (x + 1) - 2 (3 x^2 + 6x + 1)}{(x + 1)^3}$$

$$= \frac{6 x^2 + 6x + 6x + 6 - 6 x^2 - 12x - 2}{(x + 1)^3} \qquad \frac{d^2y}{dx^2} = \frac{4}{(x - 11)^3}$$

(ii) $y = e^x + \sin x, \dfrac{dy}{dx} = e^x + \cos x \qquad \dfrac{d^2y}{dx^2} = e^x - \sin x.$

(iii) $y = \dfrac{e^{-x}}{\cos 2x}, \dfrac{dy}{dx} = \dfrac{-e^{-x} \cos 2x - e^{-x} (-2 \sin 2x)}{\cos^2 2x}$

$$\frac{dy}{dx} = \frac{2^{-x} e \sin 2x - e^{-x} \cos 2x}{\cos^2 2x}$$

$$\frac{d^2y}{dx^2} = \frac{(-2 e^{-x} \sin 2x + 4 e^{-x} \cos 2x + e^{-x} \cos 2x + e^{-x} \sin 2x) \text{ c}}{\cos^4 2x}$$

$$\frac{dy}{dx} = \text{sech } t \text{ coth } t$$

$$\frac{d^2y}{dx^2} = (\text{sech } t \text{ tanh } t \text{ coth } t + \text{sech } t \, (-\text{cosech}^2 t) \times \frac{dt}{dx}$$

$$= (\text{sech } t - \text{sech } t \text{ cosech}^2 t) \quad \frac{1}{1 - \text{sech}^2 t}$$

$$= (\sec t - \text{sech } t \text{ cosech}^2 t) \quad \text{coth}^2 t$$

$$= \text{sech } t \, (1 - \text{cosech}^2 t) \text{ coth}^2 t \quad 1 - \text{coth}^2 t = -\text{cosech}^2 t$$

$$\frac{d^2y}{dx^2} = \text{sech } t \, (1 - \text{cosech}^2 t)(1 + \text{cosech}^2 t) \qquad = \text{sech } t \, (1 - \text{cosech}^4 t)$$

$$\left(\frac{dy}{dx}\right)^3 - \frac{d^2y}{dx^2} + \frac{y}{x} = \text{sech}^3 t \text{ coth}^3 - \text{sech } t + \text{cosech}^4 t + \frac{\text{sech } t}{t - \text{tanh } t}$$

189

13. $x = 3 \cos t$ $dx/dt = -3 \sin t$ $y = \cos 2t$ $dy/dt = -2 \sin 2t$

(i) $$\frac{dy}{dx} = \frac{dy/dt}{dx/dt} = \frac{-2 \sin 2t}{-3 \sin t} = \frac{2}{3} \times \frac{2 \sin t \cos t}{\sin t}$$

$$= \frac{4}{3} \cos t$$

(ii) $$\frac{d^2y}{dx^2} = -\frac{4}{3} \sin t \qquad \frac{dy}{dx} = -\frac{4}{3} \sin t \, \frac{1}{(-3 \sin t)} = \frac{4}{9}.$$

14. $x = 3 \sin 2t \, (1 - \cos 2t)$

(i) $$\frac{dx}{dt} = 6 \cos 2t \, (1 - \cos 2t) + 3 \sin 2t \, (2 \sin 2t)$$

$= 6 \cos 2t - 6 \cos^2 2t + 6 \sin^2 2t = 6 \cos 2t + 6 \, (\sin^2 2t - \cos^2 2t)$
$= 6 \cos 2t - 6 \cos^2 4t \qquad\qquad y = 3 \cos 2t \, (1 + \cos 2t)$

$$\frac{dy}{dt} = -6 \sin 2t \, (1 + \cos 2t) - 6 \cos 2t \sin 2t$$

$$= -6 \sin 2t - 6 \sin 2t \cos 2t - 6 \cos 2t \sin 2t = -6 \sin 2t - 6 \sin 4t$$

$$\frac{dy}{dx} = \frac{dy/dt}{dx/dt} = \frac{-6 \, (\sin 2t + \sin 4t)}{6 \, (\cos 2t - \cos^2 4t} \qquad \frac{dy}{dx} = -\frac{\sin 2t + \sin 4t}{\cos 2t - \cos^2 4t}.$$

15. $x = 2 \cos \Theta - \cos 2\Theta$

$$\frac{dx}{d\Theta} = -2 \sin \Theta + 2 \sin 2\Theta$$

$y = 2 \sin \Theta - \sin 2\Theta$

$$\frac{dy}{d\Theta} = 2 \cos \Theta - 2 \cos 2\Theta$$

$$\frac{dy}{dx} = \frac{\cos \Theta - \cos 2\Theta}{\sin 2\Theta - \sin \Theta} = \frac{-2 \sin 3\Theta/2 \, \sin (-\Theta/2)}{2 \cos 3\Theta/2 \, \sin \Theta/2} = \tan 3\Theta/2.$$

$$\frac{d^2y}{dx^2} = \frac{3}{2} \sec^2 \frac{3\Theta}{2} \, \frac{d\Theta}{dx} = \frac{3}{2} \sec^2 \frac{3}{2} \Theta \times \frac{1}{2 \, (\sin 2\Theta - \sin \Theta)}$$

$$= \frac{3}{4} \sec^2 \frac{3}{2} \Theta \times \frac{1}{2 \cos 3\Theta/2 \; \sin \Theta/2} = \frac{3}{8} \sec^3 \frac{3\Theta}{2} \; \text{cosec} \; \frac{\Theta}{2}.$$

16. $x = t \sin t - \cos t - 1$

$$\frac{dx}{dt} = \sin t + t \cos t + \sin t = 2 \sin t + t \cos t$$

$$y = \sin t - t \cos t \qquad \frac{dy}{dx} = \cos t - \cos t + t \sin t = t \sin t$$

$$\frac{dy}{dx} = \frac{t \sin t}{2 \sin t + t \cos t} \quad \text{at } t = \pi/2$$

$$\frac{dy}{dx} = \frac{\pi}{2} \; \frac{\sin \pi/2}{2 \sin \pi/2 + \pi/2 \; \cos \pi/2} = \frac{\pi}{4}.$$

17. $x = 3 (\Theta - \sin \Theta)$ $\qquad\qquad\qquad$ $y = 3 (1 - \cos \Theta)$

$$\frac{dx}{d\Theta} = 3 - 3 \cos \Theta \qquad\qquad\qquad \frac{dy}{d\Theta} = 3 \sin \Theta$$

$$\frac{dy}{dx} = \frac{3 \sin \Theta}{3 (1 - \cos \Theta)} = \frac{2 \sin \Theta/2 \; \cos \Theta/2}{1 - 2 \cos^2 \Theta/2 + 1}$$

$$= \frac{2 \sin \Theta/2 \; \cos \Theta/2}{2 (1 - \cos^2 \Theta/2)} = \frac{\sin \Theta/2 \; \cos \Theta/2}{\sin^2 \Theta/2} \qquad \frac{dy}{dx} = \cot \Theta/2$$

$$\frac{d^2y}{dx^2} = - \frac{1}{2} \; \text{cosec}^2 \; \Theta/2 \; \frac{d\Theta}{dx} = - \frac{1}{2} \; \text{cosec}^2 \; (\Theta/2). \; \frac{1}{3 (1 - \cos \Theta)}$$

$$= - \frac{1}{6} \; \frac{\text{cosec}^2 \; \Theta/2}{2 (1 - \cos^2 \Theta/2)} = - \frac{1}{12} \; \text{cosec}^4 \; \Theta/2.$$

18. $x = t - \tanh t$ $\qquad\qquad\qquad$ $y = \text{sech } t$

$$\frac{dx}{dt} = 1 - \text{sech}^2 t \qquad\qquad\qquad \frac{dy}{dt} = \text{sech } t \tanh t$$

$$\frac{dy}{dx} = \frac{\text{sech } t \tanh t}{1 - \text{sech}^2 t} = \frac{\text{sech } t \tanh t}{\tanh^2 t} = \frac{\text{sech } t}{\tanh t}$$

where $1 - \tanh^2 = \text{sech}^2 t$

$$-\frac{(2\,e^{-x}\sin 2x - e^{-x}\cos 2x)}{\cos^4 2x}\,2 \times 2\cos 2x\,(-\sin 2x)$$

$$\frac{d^2y}{dx^2} = \frac{5\,e^{-x}\cos 2x - e^{-x}\sin 2x}{\cos^2 2x} + \frac{4\sin 2x\cos 2x\,(2\,e^{-x}\sin 2x - e^{-x}\cos 2x)}{\cos^4 2x}$$

$$= \frac{5\,e^{-x}\cos 2x - e^{-x}\sin 2x}{\cos^2 2x} + \frac{8\,e^{-x}\sin^2 2x - 4\,e^{-x}\sin 2x\cos 2x}{\cos^3 2x}$$

(iv) $y = 3\sin 2x - 5\cos 2x$ $\dfrac{dy}{dx} = 6\cos 2x + 10\sin 2x$

$$\frac{d^2y}{dx} = -12\sin 2x + 20\cos 2x$$

(v) $y = \sin^2 x,\ \dfrac{dy}{dx} = 2\sin x\cos x = \sin 2x$ $\dfrac{d^2y}{dx^2} = 2\cos 2x$

3.   (i) $x = 2\,t^3 - 3\,t^2 + 4t + 5$ $\dfrac{dx}{dt} = 6\,t^2 - 6t + 4$, at $t = 1$

$$v = \frac{dx}{dt} = 6 - 6 + 4 = 4 \text{ m/s}$$

(ii) $v = 4$ m/s when $t = 0$

(iii)   at $t = 5$, $v = 6\,(5)^2 - 6\,(5) + 4 = 150 - 30 + 4$ $v = 124$ m/s

(iv) $a = \dfrac{dv}{dt} = 12t - 6$, at $t = 0$, $a = 6$ m/s^2

at $t = 2$, $a = 24 - 6 = 18$ m/s^2.

(v)   at $t = 2$, $x = 2\,(2)^3 - 3\,(2)^2 + 4\,(2) + 5 = 16 - 12 + 8 + 5$
$x = 17$ m; at $t = 3$, $x = 2\,(3)^3 - 3\,(3)^2 + 4\,(3) + 5$
$x = 44$ m; at $t = 5$, $x = 2\,(5)^3 - 3\,(5)^2 + 4\,(5) + 5$
$x = 200$ m.

4. $s = 30\,t^2 - 3t + 5$

(i) $\dfrac{ds}{dt} = 60t - 3$ at $t = 3$, $v = 60\,(3) - 3 = 177$ m/s

(ii) $a = \dfrac{d^2s}{dt^2} = \dfrac{dv}{dt} = 60$ m/s^2

(iii) $v = 0$ at rest, $0 = 60t, t = 1/20$

$$s = 30 \, (1/20)^2 - 3 \, (1/20) + 5 = \frac{30}{400} - \frac{3}{20} + 5 = \frac{30 - 60 + 2000}{400}$$

$$s = \frac{1970}{400} = 4.925 \text{ m.}$$

(iv) $v = 57$ m/s, $57 = 60t - 3$, $60 = 60t, t = 1$s.

(v) $v = 60 \, (10) - 3 = 600 - 3 = 597$ m/s.

5. $s = 20 \, t^2$

(i) $\dfrac{ds}{dt} = 40t$ \qquad\qquad (ii) $\dfrac{ds}{dt} = 40$ m/s

(iii) $s = 1500 = 20 \, t^2,$ \qquad $t = \sqrt{\dfrac{1500}{20}} = 8.66$ s

(iv) $\dfrac{d^2 s}{dt^2} = 40$ m/s ^{2}.

SOLUTIONS 8

1. $x = a (2\Theta - \sin 2\Theta)$ and $y = a (1 - \cos 2\Theta)$

$$\frac{dx}{d\Theta} = 2a - 2a \cos 2\Theta \text{ and } \frac{dy}{d\Theta} = 2a \sin 2\Theta$$

$$\frac{dy/d\theta}{dx/d\theta} = \frac{2a \sin 2\theta}{2a - 2a \cos 2\theta} = \frac{\sin 2\theta}{1 - \cos 2\theta} = \frac{2 \sin \theta \cos \theta}{1 - 2\cos^2 \theta + 1}$$

$$= \frac{\sin \Theta \cos \Theta}{1 - \cos^2 \Theta} = \frac{\cos \Theta}{\sin \Theta} = \cot \Theta$$

$y = mx + c$ the equation of the tangent at
$\Theta = \pi/4$, $x = a (2\pi/4 - \sin 2\pi/4) = a (\pi/2 - \sin \pi/2)$
         $x = a (\pi/2 - 1)$
         $y = a (1 - \cos 2 \pi/4) = a (1 - \cos \pi/2)$ $y = a$
$P (a (a (\pi/2 - 1), a)$ $y = \cot \pi/4$ $x + c = x + c$
$a = \cot \pi/4 (a (\pi/2 - 1)) + c$

$c = a - a \pi/2 + a = 2a - a \pi/2$

$$\boxed{y = x + 2a - a \pi/2}$$ Equation of tangent

$y = - \tan \pi/4$ $x + c = - x + c$ $a = - a \pi/2 + a + c, c = a \pi/2$

$$\boxed{y = x + a \pi/2}$$ Equation of normal.

2. $x = 2 \cos t - \cos 2t$ $y = 2 \sin t - \sin 2t$

$$\frac{dx}{dt} = - 2 \sin t + 2 \sin 2 t \qquad \frac{dy}{dt} = 2 \cos t - 2 \cos 2 t$$

$$\frac{dy}{dx} = \frac{dy/dt}{dx/dt} = \frac{2 (\cos t - \cos 2 t)}{2 (\sin 2 t - \sin t)} = \frac{- 2 \sin 3t/2 \sin (- t/2)}{2 \cos 3t/2 \sin t/2} = \tan 3t/2$$

The equation of the tangent $y = mx + c$ $2 \sin t - \sin 2t$
$= \tan 3t/2 (2 \cos t - \cos 2t) + c$

$c = 2 \sin t - \sin 2t - 2 \cos t \tan 3t/2 - \cos 2t \tan 3t/2$

$y = (\tan 3 t/2) x + 2 \sin t - \sin 2t - 2 \cos t \tan 3t/2 - \cos 2t \tan 3 t/2.$

The gradient of the normal is found from $m_1 \, m_2 = -1$.

If $m_1 = \tan 3 t/2$, then $m_2 = - \cot 3 t/2$.

The equation of the normal

$$y = (- \cot 3 t/2) \, x + c$$

$2 \sin t - \sin 2t = - \cot 3 t/2 \, (2 \cot t - \cos 2 t) + c$

$c = 2 \sin t - \sin 2t + 2 \cos t \cot 3t/2 + \cos 2t \cot 3t/2$

$y = (- \cot 3t/2) \, x + 2 \sin t - \sin 2t + 2 \cos t \cot 3t/2 + \cos 2t \cot 3t/2.$

3. $x = 2 \cos \Theta$ $y = 3 \sin \Theta$

$\dfrac{dx}{d\theta} = -2 \sin \theta$ $\dfrac{dy}{d\theta} = 3 \cos \theta$

$\dfrac{dy/d\theta}{dx/d\theta} = \dfrac{dy}{dx} = -\dfrac{3}{2} \cot \Theta$ the gradient of the tangent, the gradient of

the normal is $\dfrac{2}{3} \tan \theta$.   The equation of the normal $y = \left(\dfrac{2}{3} \tan \theta \right) x + c$

$3 \sin \Theta = \left(\dfrac{2}{3} \tan \Theta \right) (2 \cos \Theta) + c$

$c = 3 \sin \Theta - \dfrac{4}{3} \sin \Theta = \dfrac{5}{3} \sin \Theta$

$$\boxed{y = \dfrac{2}{3} \tan \Theta \, x + \dfrac{5}{3} \sin \Theta}$$

4. $x = 5 \sin^3 \Theta, \quad \dfrac{dx}{d\theta} = 15 \sin^2 \theta \cos \theta, \quad \dfrac{5}{8} = 5 \sin^3 \theta, \sin \theta = \dfrac{1}{2}$

$\Theta = \pi/6$, taking the principal value.

$y = 5 \cos^3 \Theta, \quad \dfrac{dy}{d\theta} = -15 \cos^2 \Theta \sin \Theta; \, y = 5 \cos^3 \Theta = 5 \left(\dfrac{\sqrt{3}}{2} \right)^3$

$y = \dfrac{15}{8} \sqrt{13}.$

$\dfrac{dy/d\theta}{dx/d\theta} = \dfrac{dy}{dx} = \tan \Theta$

The equation of the tangent is $y = (\tan \Theta) x + c$

$5 \cos^3 \Theta = (5 \sin^3 \Theta) \tan \Theta + c \quad c = 5 \cos^3 \Theta - 5 \sin^3 \Theta \tan \Theta$

$y = (\tan \Theta) x + 5 \cos^3 \Theta = 5 \sin^3 \Theta \tan \Theta.$

$$y = \frac{1}{\sqrt{3}} x + \frac{15}{8} \sqrt{3} - \frac{5}{8\sqrt{3}} \frac{\sqrt{3}}{\sqrt{3}} = \frac{1}{\sqrt{3}} x + \frac{5}{3} \sqrt{3}.$$

The equation of the normal is

$y = - \cot \Theta \, x + c \quad 5 \cos^3 \Theta = - 5 \sin^3 \Theta \cot + c$

$c = 5 \cos^3 \Theta + 5 \sin^3 \Theta \cot \Theta = 5 \cos^3 \Theta + 5 \sin^2 \Theta \cos \Theta$

$y = - x \cot \Theta + 5 \cos^3 \Theta + 5 \sin^2 \Theta \cos \Theta.$

$$y = - \sqrt{3} \, x + \frac{15}{8} \sqrt{3} + 5 \times \frac{1}{4} \frac{\sqrt{3}}{2} = - \sqrt{3} \, x + \frac{5}{2} \sqrt{3}.$$

5. $x^2 + y^2 - x - y - 2 = 0$ differentiating with respect to x

$$2x + 2y \frac{dy}{dx} - 1 - \frac{dy}{dx} = 0, \quad \frac{dy}{dx} = \frac{1 - 2x}{2y - 1}.$$

At the point $(1, -1)$, $\dfrac{dy}{dx} = \dfrac{1 - 2}{- 2 - 1} = \dfrac{- 1}{- 3} = \dfrac{1}{3}$,

at the point $(1, 2)$, $\dfrac{dy}{dx} = \dfrac{1 - 2}{4 - 1} = - \dfrac{1}{3}.$

The equations of the tangents are $y = \dfrac{1}{3} x + c,$

and $y = - \dfrac{1}{3} x + c_2$ respectively.

To find $c, x = 1, y = -1, -1\dfrac{1}{3} = c_1, c_1 = - \dfrac{4}{3}$

$y = \dfrac{1}{3}x - \dfrac{4}{3}$ $\boxed{3y + 4 = x}$

The find $c_2, x = 1, y = 2, \quad x = 1, y = 2, 2 = - \dfrac{1}{3} + c_2, c_2 = 7/3$

$y = - \dfrac{1}{3}x + \dfrac{7}{3},$ $\boxed{3y + x = 7}$

6. $y^2 = 4x$, differentiating with respect to x,

$$2y \frac{dy}{dx} = 4, \qquad y \frac{dy}{dx} = 2$$

at $x = 1, y = -2, -2 \frac{dy}{dx} = 2, \qquad \frac{dy}{dx} = -1.$

The equation of the normal, $y = x + c - 2 = 1 + c, c = -3,$ $\boxed{y = x - 3}$

7. $xy = 9$ differentiating with respect to x

$$y + x \frac{dy}{dx} = 0, y = -9, x = -1, -9 - \frac{dy}{dx} = 0$$

$\frac{dy}{dx} = -9,$ the equation of the tangent

$$y = -9x + c, -9 = 9 + c, \quad c = -18$$

$$\boxed{y = -9x - 18}$$

The equation of the normal

$$y = \frac{1}{9} x + c, \qquad -9 = -\frac{1}{9} + c, \qquad c = -8 \frac{8}{9}$$

$$y = \frac{1}{9} x^2 - \frac{80}{9}, \qquad \boxed{9y + 80 = x}$$

SOLUTIONS 9

1. V = volume of the trough = area of trapezium x length of trough

$$= \frac{1}{2} (a + b) h l$$

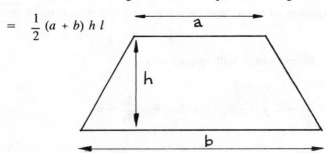

Fig. 42

$$V = \frac{1}{2} (1 + 2) h \, 5 \qquad \frac{dV}{dh} = \frac{15}{2} \qquad \frac{dV/dt}{dh/dt} = 7.5$$

$$dh/dt = \frac{dV/dt}{7.5} = \frac{1 \times 10^{-3}}{7.5} = \frac{2}{15} \times 10^{-3} \text{ m/s}$$

2. $V = \frac{4}{3} \pi r^3$ volume of spherical balloon

$$S = \frac{dV}{dr} = 4\pi r^2 = \text{ surface area}$$

$$\frac{dV/dt}{dr/dr} = 4 \pi r^2, \, dr/dt = \frac{dV}{dt} \times \frac{1}{4 \pi r^2} = \frac{5 \times 10^{-3}}{4 \pi \, 10^2}$$

$$= 3.98 \times 10^{-6} \text{ m/s}$$

$$\frac{dS}{dr} = 8\pi r, \, \frac{dS/dt}{dr/dt} = 8\pi r \qquad dS/dt = 8\pi r \, (3.98 \times 10^{-6}) = 1 \times 10^{-3} \text{ m}^2/\text{s}$$

3. $V = \frac{4}{3} \pi r^3$

$$S = \frac{dV}{dr} = 4\pi r^2 \qquad \frac{dV/dt}{dr/dt} = 4\pi r^2 \qquad \frac{dS}{dr} = 8\pi r$$

$$\frac{dS/dt}{dr/dt} = 8\pi r \qquad dS/dt = 8\pi r \frac{dr}{dt} \qquad dr/dt = \frac{dS/dt}{8\pi r} = \frac{1}{8\pi \, 100}$$

$$\frac{dS}{dt} = 1 \ \text{mm}^2/s \quad dV/dt = 4\pi \ 100^2 \ \frac{1}{\pi \ 800} = \frac{4 \times 10{,}000}{800} = 50 \ \frac{\text{mm}^2}{5}$$

4. $\qquad V = \dfrac{4}{3} \ \pi \ r^3, \dfrac{dV}{dr} = 4\pi \ r^2 = S, \qquad \dfrac{ds}{dr} = 8\pi r$

$$\frac{dV/dt}{dr/dt} = 4\pi \ r^2 = \frac{100}{dr/dt} \frac{dr}{dt} = \frac{100}{4 \ \pi \times 75^2} = 1.42 \times 10^{-3} \ \text{mm/s}$$

5. $\qquad \dfrac{dV}{dh} = 3 \ e^{3h} + 5 \ e^h + 1 \ \dfrac{dV/dt}{dh/dt} = 3 \ e^{3h} + 5 \ e^h + 1$

$$dV/dt = (3 \ e^{3h} + 5 \ e^h + 1) \ dh/dt$$

$$= (3 \ e^3 + 5 \ e^1 + 1) \ 0.03 = 2.25 \ \text{m}^3/s$$

6. $\qquad 4 \ x^2 + 9 \ y^2 = 36$ differentiating with respect to t

$$8 \ \frac{dx}{dt} + 18 \ \frac{dy}{dr} = 0$$

$$\frac{dy}{dt} = -\frac{4}{9} \frac{dx}{dt} = \frac{4}{9} \ (0.1) = -\frac{0.4}{9} = -0.044 \ \text{cm/s}$$

the rate of decrease is 0.044 cm/s.

7. $\qquad V = \dfrac{\pi \ h^2}{3} \ (3r - h) \qquad\qquad \dfrac{dV}{dr} = \dfrac{\pi \ h^2}{3} \ 3 = \pi \ h^2$

$$\frac{dV/dt}{dr/dt} = \pi \ 20^2$$

$$\frac{dV}{dt} = 400 \ \pi \times 5 = 2000 \ \pi \ \text{cm}^3/s = 6.28 \times 10^3 \text{cm}^3/s$$

8. $\qquad A = \sqrt{s \ (s - a) \ (s - b) \ (s - c)} = \sqrt{15 \times 5 \times 5 \times 5} = 25 \ \sqrt{3} \ \text{cm}^2$

where $\quad s = \dfrac{10 + 10 + 10}{2} = 15 \ \text{cm} =$ the semi-perimeter

$$s = \frac{3}{2} \ a, \qquad a = \frac{2}{3} \ s$$

$$A = \sqrt{s\left(s - \frac{2}{3}\,s\right)^3} = s^2\sqrt{\frac{1}{3^3}} = \frac{s^2}{3\sqrt{3}}$$

$$\frac{dA}{dt} = \frac{2s}{3\sqrt{3}} \qquad\qquad dA/dt = \frac{2s}{3\sqrt{3}}\,ds/dt$$

$$= \frac{2 \times 15}{3\sqrt{3}} \times \frac{1}{2} = \frac{5}{\sqrt{3}}\,\frac{\sqrt{3}}{\sqrt{3}} = \frac{5}{3}\sqrt{3}\ \text{cm}^2/\text{s}$$

$$A = \frac{1}{2}\,a^2\sin 60° \qquad\qquad \frac{dA}{da} = a\sin 60°$$

$$\delta A = a\sin 60°\ (\delta a) = 10 \times 0.866 \times 0.1 = 0.866\ \text{cm}^2.$$

9.   (i) $\displaystyle\lim_{x \to 0} \frac{\sin 5x}{5x} = \frac{0}{0}$ indeterminate

applying L'Hôpital's rule

$$\lim_{x \to 0} \frac{5\cos 5x}{5} = \frac{5}{5}\cos 5\,(0) = \frac{5}{5}\,(1) = 1.$$

(ii) $\displaystyle\lim_{x \to 0} \frac{\tan kx}{x} = \frac{0}{0}$ indeterminate

applying L'Hôpital's rule

$$\lim_{x \to 0} \frac{k\sec^2 kx}{1} = k$$

(iii) $\displaystyle\lim_{x \to 0} \frac{2\sin^{-1} x}{3x} = \frac{0}{0}$ indeterminate

applying L'Hôpital's rule

$$\lim_{x \to 0} \frac{2}{3}\,\frac{1}{\sqrt{1 - x^2}} = \frac{2}{3}.$$

(iv)
$$\lim_{x \to 0} \frac{1 - \cos x}{x^2} = \frac{0}{0} \quad \text{indeterminate}$$

applying L'Hôpital's rule

$$\lim_{x \to 0} \frac{\sin x}{2x} = \frac{0}{0} \quad \text{determinate}$$

applying L'Hôpital's rule

$$\lim_{x \to 0} \frac{\cos x}{2} = \frac{1}{2}.$$

(v)
$$\lim_{x \to 0} \left(\frac{1}{\sin x} - \frac{1}{\tan x} \right) = \lim_{x \to 0} \left(\frac{\tan x - \sin x}{\sin x \tan x} \right) = \frac{0}{0}$$

indeterminate

applying L'Hôpital's rule

$$\lim_{x \to 0} \frac{\sec^2 x - \cos x}{\cos x \tan x + \sin x \sec^2 x} = \frac{0}{0} \quad \text{indeterminate}$$

$$\lim_{x \to 0} \frac{1 - \cos^3 x}{\cos^3 x \tan x + \sin x} = \frac{0}{0} \quad \text{indeterminate}$$

applying the rule again

$$\lim_{x \to 0} \frac{3 \cos^2 x \sin x}{3 \cos^2 x (- \sin x) \tan x + \cos^3 x \sec^2 x + \cos x} = 0.$$

Alternatively

$$\lim_{x \to 0} \left(\frac{1}{\sin x} - \frac{1}{\tan x} \right) = \lim_{x \to 0} \left(\frac{1 - \cos x}{\sin x} \right) = \frac{0}{0}$$

indeterminate applying the rule $\lim\limits_{x \to 0} \dfrac{\sin x}{\cos x} = 0.$

(vi) $\lim\limits_{x \to \pi/2} \dfrac{\cos x}{\sqrt{1 - \sin x^{2/3}}} = \dfrac{0}{0}$ indeterminate

applying the rule

$$\lim_{x \to \pi/2} \frac{- \sin x}{\frac{1}{3} (1 - \sin x)^{-2/3} (- \cos x)} = \frac{0}{-1/3} = 0.$$

(vii) $\lim\limits_{x \to 0} \dfrac{1 - \cos^3 x}{x \sin 2x} = \dfrac{0}{0}$ indeterminate

applying the rule

$$\lim_{x \to 0} \frac{3 \cos^2 x \sin x}{\sin 2x + 2x \cos 2x} = \frac{0}{0} \text{ indeterminate}$$

$$\lim_{x \to 0} \frac{- 6 \cos x \sin^2 x + 3 \cos^3 x}{2 \cos 2x + 2 \cos 2x - 4x \sin 2x} = \frac{3}{4}.$$

(viii) $\lim\limits_{x \to \pi/2} \dfrac{1 - \sin x}{\left(\dfrac{\pi}{2} - x \right)^2} = \dfrac{0}{0}$ indeterminate

202

applying the rule

$$\lim_{x \to \pi/2} \quad \frac{-\cos x}{2\,(\pi/2 - x)\,(-1)} = \frac{0}{0} \quad \text{indeterminate}$$

applying the rule $\lim_{x \to \pi/2} \quad \dfrac{\sin x}{2} = \dfrac{1}{2}.$

(ix) $\lim_{x \to 0} \quad \dfrac{\sin 3x}{\sin 2x} = \dfrac{0}{0} \quad \text{indeterminate}$

applying the rule

$$\lim_{x \to 0} \quad \frac{3 \cos 3x}{2 \cos 2x} = \frac{-3}{2} = -\frac{3}{2}.$$

(x) $\lim_{x \to 0} \quad \dfrac{e^x - e}{x - 1} = \dfrac{0}{0} \quad \text{indeterminate}, \quad \lim_{x \to 0} \quad \dfrac{e^x}{1} = 1.$

(xi) $\lim_{x \to 0} \quad \dfrac{e^x - \cos x}{x^2} = \dfrac{1 - 1}{0} = \dfrac{0}{0} \quad \text{indeterminate}$

applying the rule

$$\lim_{x \to 0} \quad \frac{2x\,e^{x^2} + \sin x}{2x} = \frac{0}{0} \quad \text{indeterminate}$$

$$\lim_{x \to 0} \quad \frac{2\,e^{x^2} + 4\,x^2\,e^{x^2} + \cos x}{2} = \frac{2 + 0 + 1}{2} = \frac{3}{2}.$$

(xii) $\lim_{x \to 0} \quad \dfrac{e^x - e^{-x}}{\sin x} = \dfrac{1 - 1}{0} = \dfrac{0}{0} \quad \text{indeterminate}$

applying the rule

$$\lim_{x \to 0} \quad \frac{e^x + e^{-x}}{\cos x} = \frac{2}{1} = 2.$$

(xiii) $\lim_{x \to 0} \quad \dfrac{\ln \cos x}{x^2} = \dfrac{0}{0}$ indeterminate

applying the rule

$$\lim_{x \to 0} \quad \frac{-\sin x / \cos x}{2x} = \frac{0}{0} \text{ indeterminate applying the rule}$$

$$\lim_{x \to 0} \quad \frac{-\sec^2 x}{2} = -\frac{1}{2}.$$

(xiv) $\lim_{x \to 0} \quad \dfrac{1 - \cos (1 - \cos x)}{x^3} = \dfrac{0}{0}$ indeterminate

applying the rule

$$\lim_{x \to 0} \quad \frac{\left[\sin (1 - \cos x) \right] (\sin x)}{3 x^2} = \frac{0}{0} \text{ indeterminate}$$

applying the rule again

$$\lim_{x \to 0} \quad \frac{\left[(\cos (1 - \cos x) \right] \sin^2 x + \left[\sin (1 - \cos x) \right] \cos x}{6x} = \frac{0}{0}$$

indeterminate

$$\lim_{x \to 0} \frac{[-\sin(1-\cos x)]\sin^3 x + [\cos(1-\cos x)]2\sin x \cos x + \cos(1-\cos x)\sin x \cos x + \sin(1-\cos x)(-\sin x)}{6}$$

$$= 0$$

(xv) $\quad \lim_{x \to 0} \dfrac{1 - \sin(1 + \sin x)}{x^4} = \dfrac{1 - \sin 1}{x^4} = \infty.$

REMEMBER THAT L'HÔPITAL'S RULE IS APPLIED ONLY IF THE RESULT IS INDETERMINATE SUCH AS $\dfrac{0}{0}$ OR $\dfrac{\infty}{\infty}$.

SOLUTIONS 10

1.   (i) $y = x^3 + 3x + 3$

$\dfrac{dy}{dx} = 3x^2 + 3$, this cannot be equal to zero, the gradient is every

where positive, and therefore there are no turning points.

$\dfrac{d^2y}{dx^2} = 6x = 0$, at $x = 0$ there is a point of inflexion.

If $x = 0$, $y = 3$ and the sketch of the graph is as shown.

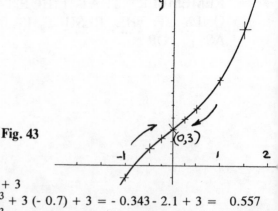

Fig. 43

$f(x) = x^3 + 3x + 3$
$f(-0.7) = -0.7^3 + 3(-0.7) + 3 = -0.343 - 2.1 + 3 = 0.557$
$f(-0.9) = -0.9^3 + 3(-0.9) + 3 = -0.729 - 2.7 + 3 = -0.429$
The root of $f(x) = 0$ lies between -0.9 and -0.7.

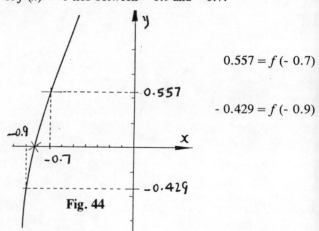

$0.557 = f(-0.7)$

$-0.429 = f(-0.9)$

Fig. 44

(ii) $y = x^3 + 3x + 28 = 0$

$\dfrac{dy}{dx} = 3x^2 + 3$ this cannot be equal to zero, the gradient is always

206

positive and therefore there are no turning points.

$\dfrac{d^2y}{dx^2} = 6x = 0$, at $x = 0$ there is a point of inflexion.

If $x = 0, y = 28$ and the sketch is

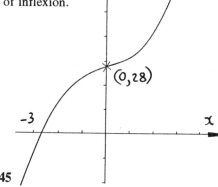

Fig. 45

$f(-2.6) = (-2.6)^3 + 3(-2.6) + 28 = -17.576 - 7.8 + 28$
$\qquad = 2.624$

$f(-2.8) = (-2.8)^3 + 3(-2.8) + 28 = -21.952 - 8.4 + 28$
$\qquad = 2.352$

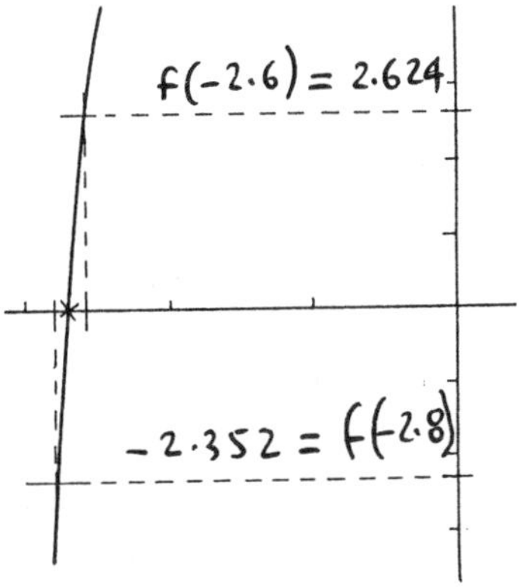

Fig. 46

$f(-2.6) = 2.624$

$-2.352 = f(-2.8)$

The root for $f(x) = 0$ lies between -2.8 and -2.6.

$$x_{n+1} = x_n - \dfrac{f(x_n)}{f'(x_n)}$$

(i) $f(x) = x^3 + 3x + 3$
$f'(x) = 3x^2 + 3$
$f(x_n) = f(-0.75) = -0.75^3 + 3(-0.75) + 3 = +0.328125$
$f'(x_n) = f'(-0.75) = 3(-0.75)^2 + 3 = 4.6875$

$x_{n+1} = -0.75 - \dfrac{(0.328125)}{4.6875} = -0.82$, to three significant figures.

(ii) $f(x) = x^3 + 3x + 28$
$f'(x) = 3x^2 + 3$
$f(x_n) = f(-2.7) = -2.7^3 + 3(-2.7) + 28 = 0.217$
$f'(-2.7) = 3(-2.7)^2 + 3 = 24.87$

$$x_{n+1} = -2.7 - \frac{0.217}{24.87} = -2.7087254 = -2.71 \text{ to three significant figures.}$$

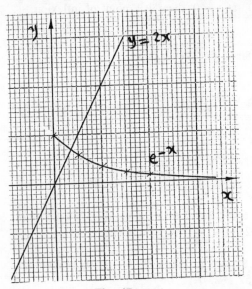

Fig. 47

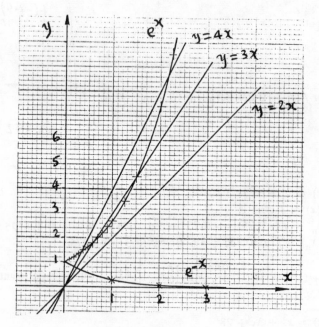

Fig. 48

2.   The graphs of $y = 2x$ and $y = -e^{-x}$ are plotted.

x	0	0.5	1	1.5	2
e^{-x}	1	0.607	0.368	0.223	0.135

The intersection of the two graphs is $x = 0.35$.

$f(x) = 2x - e^{-x}$

$f^{I}(x)\ 2 + e^{-x}$

$f(x_n) = f(0.35) = 2(0.35) = e^{-0.35} = 0.7 - 0.704688$

$= -0.004688$

$f^{I}(x_n) = f^{I}(0.35) = 2 + e^{-0.35} = 2.7046881$

$$x_{n+1} = x_n - \frac{f(x_n)}{(f^{I}(x_n)} = 0.35 - \frac{(-0.004688)}{2.7046881} = 0.351733$$

$= 0.3517$ to four decimal places.

$f(x) = 3x - e^x$

$y = 3x, y = e^x$.

These graphs are plotted giving two intersections at $x = 0.65$ and $x = 1.60$.

$f^{I}(x) = 3 - e^x$

$f(x_n) = f(0.65) = 3(0.65) - e^{0.65} = 1.95 - 1.9155478 = 0.0345$

$f^{I}(x_n) = f^{I}(0.65) = 3 - e^{0.65} = 1.0845$

$$x_{n+1} = x_n - \frac{f(x_n)}{f^{I}(x_n)} = 0.65 - \frac{0.345}{1.0845} = 0.618$$

$f(x_n) = f(1.6) = 3(1.6) - e^{1.6} = -0.153$

$f^{I}(x_n) = f^{I}(1.6) = 3 - e^{1.6} = -1.953$

$$x_{n+1} = x_n - \frac{f(x_n)}{f^{I}(x_n)} = 1.60 - \frac{(0.153)}{(-1.953)} = 1.522$$

$= 4x - e^x$ $y = 4x$ and $y = e^x$.

These graphs are plotted giving two intersections at $x = 0.4$ and $x = 2.1$.

$f(x) = 4x - e^x, f(0.4) = 1.6 = e^{0.04} = 0.1881753$

$f^I(x) = 4 - e^x, f^I(0.4) = 4 - e^{0.4} = 2.5081753$

$$x_{n+1} = 0.4 - \frac{0.1881753}{2.5081753} = 0.325$$

$f(2.1) = 4(2.1) - e^{2.1} = 8.4 - 8.1661699 = 0.23383$

$f^I(2.1) = 4 - e^{2.1} = 8.1661699$

$$x_{n+1} = 2.1 - \frac{0.23383}{8.1661699} = 2.071366 \qquad x_{n+1} = 2.07.$$

3. $\quad f(x) = \cos\left(2x - \frac{2\pi}{3}\right) - \frac{1}{2}x \quad f(1.414^c) = \cos\left(2.828 - \frac{2\pi}{3}\right) - \frac{1}{2}(1.4$

$$= 0.7427655 - 0.707 \qquad = 0.0357655$$

$$f^I(x) = -2\sin\left(2x - \frac{2\pi}{3}\right) - \frac{1}{2}$$

$$f^I(1.414^c) = 2\sin$$

$$x_{n+1} = x_n - \frac{f(x_n)}{f^I(x_n)} = 1.414 - \frac{0.0357655}{-1.9672098} = 1.3958$$

$$= 1.396 \text{ to three decimal places}$$

$$f(1.396) = \cos\left(2 \times 1.396 - \frac{2\pi}{2}\right) - \frac{1}{2}(1.396) = 0.0683829$$

$$f^I(1.396) = -2\sin\left(2 \times 1.396 - \underline{\quad}\right)$$

$$x_{n+2} = x_{n+1} - \frac{f(x_{n+1})}{f^I(x_{n+1})} = 1.396 - \frac{0.0683829}{-1.7847679}$$

$$= 1.434 \text{ to three decimal places.}$$

$$\frac{2\pi}{3} - \frac{1}{2} = -1.7847679$$

$$\left(2.828 - \frac{2\pi}{3}\right) - \frac{1}{2} = -1.9672098$$

SOLUTIONS 11

1. $\qquad$ differentiating with respect to x, we have

$$-\sin x = -\frac{2x}{2} + \frac{4x^3}{4!} - \frac{6x^5}{6!} + \dots \qquad \sin x = x - \frac{x^3}{3!} + \frac{x^5}{5!} - \dots$$

2. $f(x) = \sin^{-1} x = y, f(0) = 0 \quad x = \sin y$ by definition of the inverse function

and differentiating with respect to y $\quad \dfrac{dx}{dy} = \cos y$

$$\frac{dy}{dx} = \frac{1}{\cos y} = \frac{1}{\sqrt{1 - \sin^2 y}} = \frac{1}{\sqrt{1 - x^2}} = (1 - x^2)^{-1/2}$$

$$\frac{d^2y}{dx^2} = -\frac{1}{2}(1 + x^2)^{-3/2}(-2x) = (1 - x^2)^{-3/2}$$

$$\frac{d^3y}{dx^3} = (1 - x^2)^{-3/2} + x\left(-\frac{3}{2}\right)(1 - x^2)^{-5/2}(-2x)$$

$$= (1 - x^2)^{-3/2} + 3x^2(1 - x^2)^{-5/2}$$

$f'(x) = (1 - x^2)^{-1/2}, f'(0) = 1$

$f''(x) = x(1 - x^2)^{-3/2}, f''(0) = 0$

$f'''(x) = (1 - x^2)^{-3/2} + 3x^2(1 - x^2)^{-5/2}, \qquad f'''(0) = 1$

$f^{iv}(x) = -\dfrac{3}{2}\left(1 - x^2\right)^{-5/2}(-2x) + 6x\left(1 - x^2\right)^{-5/2}$

$+ 3x^2\left(-\dfrac{5}{2}\right)\left(1 - x^2\right)^{-7/2}(-2x)$

$= 3x(1 - x^{2-5/2} + 6x(1 - x^2)^{-5/2} + 30x^3(1 - x^2)^{-7/2} \quad f^{iv}(0) = 0$

$f^{v}(x) = 3(1 - x^2)^{-5/2} + 3x\left(-\dfrac{5}{2}\right)\left(1 - x^2\right)^{-7/2}(-2x) + 6\left(1 - x^2\right)^{-5/2}$

$+ 6x\left(-\dfrac{5}{2}\right)\left(1 - x^2\right)^{-7/2}(-2x)$

$$\cos x = 1 - \frac{x^2}{2!} + \frac{x^4}{4!} - \frac{x^6}{6!} + \dots$$

$+ 90 x^2 (1 - x^2)^{1- 7/2} + 30 x^3 (1 - x^2)^{- 9/2} (- 2x) (- 7/2)$

$= 3 (1 - x^2)^{- 5/2} + 15 x^2 (1 - x^2)^{- 7/2} + 6 (1 - x^2)^{5/2} + 60 x^2 (1 - x^2)^{- 7/2}$

$+ 90 x^2 (1 - x^2)^{- 7/2}$

$+ 210 x^4 (1 - x^2)^{- 9/2}, f^v (0) = 3 + 6 = 9$

$\sin^{-1} x = x + \dfrac{1}{6} x^3 + \dfrac{3}{40} x^5.$

3. $f (x) = \tan^{-1} x \ y = \tan^{-1} x$ by the definition of inverse function

       $x = \tan y$

$$\frac{dx}{dy} = \sec^2 y$$

$$\frac{dy}{dx} = \frac{1}{\sec^2 y} = \frac{1}{1 + \tan^2 y} = \frac{1}{1 + x^2}$$

$$f^{\,I} (x) = \frac{1}{1 + x^2}, f^{\,I} (x) = (1 + x^2)^{-1}$$

$f^{\,II} (x) = - (1 + x^2)^{-2} (2x) = - 2x (1 + x^2)^{-2}$
$f^{\,III} (x) = - 2 (1 + x^2)^{-2} + 4x (1 + x^2)^{-3} (2x)$
           $= - 2 (1 + x^2)^{-2} + 8 x^2 (1 + x^2)^{-3}$
$f^{\,iv} (x) = 4 (1 + x^2)^{-3} (2x) + 16x (1 + x^2)^{-3} - 24 x^2 (1 + x^2)^{-4} (2x)$
           $= 8x (1 + x^2)^{-3} + 16x (1 + x^2)^{-3} - 48 x^3 (1 + x^2)^{-4}$
$f^{\,v} (x) = 8 (1 + x^2)^{-3} - 24x (1 + x^2)^{-4} (2x)$
           $+ 16 (1 + x^2)^{-3} - 48x (1 + x^2)^{-4} 2x$
           $- 144 x^2 (1 + x^2)^{-4} - 48 x^3 (- 4) (1 + x^2)^{-5} (2x)$
           $= 8 (1 + x^2)^{-3} - 48x (1 + x^2)^{-4} + 16 (1 + x^2)^{-3} - 96 x^2 (1 + x^2)^{-4}$
           $- 144 x^2 (1 + x^2)^{-4} + 384 x^4 (1 + x^2)^{-5}$

$f (0) = 0, f^{\,I} (0) = 1, f^{\,II} (0) = 0, f^{\,III} (0) = - 2, f^{\,iv} (0) = 0$
$f^{\,v} (0) = 24$

$$\tan^{-1} x = x - \frac{1}{3} x^3 + \frac{1}{5} x^5$$

$$a_0 = 0, a_1 = 1, a_2 = 0, a_3 = - \frac{1}{3}, a_4 = 0, \quad \text{and} \quad a_5 = \frac{1}{5}.$$

4. $f(x) = \dfrac{1}{1 + x^2}$

   (a) $\dfrac{1}{1 + x^2} = (1 + x^2)^{-1} = 1 + (-1) x^2$

            $+ (-1) \dfrac{(-2)}{2!} x^4 + (-1)(-2)(-3) \dfrac{x^6}{3!} + \ldots$

            $= 1 - x^2 + x^4 - x^6 + \ldots - 1 < x^2 < 1.$

   (b) $f(x) = \dfrac{1}{1 + x^2} = (1 + x^2)^{-1} \qquad f(0) = 1$

       $f'(x) = -(1 + x^2)^{-2}(2x) \qquad f'(0) = 0$

       $f''(x) = 2(1 + x^2)^{-3}(2x)^2 - (1 + x^2)^{-2} \, 2$

            $= 8 x^2 (1 + x^2)^{-3} - 2(1 + x^2)^{-2} \qquad f''(0) = -2$

       $f'''(x) = -24 x^2 (1 + x^2)^{-4}(2x) + 16x(1 + x^2)^{-3} + 4(1 + x^2)^{-3}(2x)$

            $= -48 x^3 (1 + x^2)^{-4} + 16x(1 + x^2)^{-3} + 8x(1 + x^2)^{-3}$

       $f'''(0) = 0$

       $f^{iv}(x) = -144 x^2 (1 + x^2)^{-4} + 192 x^3 (1 + x^2)^{-5} \, 2x + 16(1 + x2)^{-3}$

           $- \quad 48x \, 2x (1 + x^2)^{-4}$

           $+ 8(1 + x^2)^{-3} - 24x(1 + x^2)^{-4}(2x)$

   $f^{iv}(0) = 16 + 8 = 24 \quad f(x) = \dfrac{1}{1 + x^2} = 1 - x^2 + x^4 - \ldots$

integrating both sides with respect to x.

$$\int \dfrac{1}{1 + x^2} \, dx = \tan^{-1} x = x - \dfrac{x^3}{3} + \dfrac{x^5}{5} - \ldots$$

$c = $ the arbitrary constant $= 0$ if $x = 0$.

5. $\displaystyle\int \dfrac{1}{(25 - x^2)} \, dx = \dfrac{1}{10} \ln \dfrac{5 + x}{5 - x}$

since $\dfrac{1}{25 - x^2} = \dfrac{1}{(5 - x)(5 + x)} \equiv \dfrac{A}{5 - x} + \dfrac{B}{5 + x}$

$$1 \equiv A (5 + x) + B (5 - x)$$

If $x = 5, A = 1/10$; and if $x = -5, B = 1/10$.

$$\int \frac{1}{25 - x^2} \, dx = \int \frac{1/10}{5 - x} \, dx + \int \frac{1/10}{5 + x} \, dx$$
$$= -\frac{1}{10} \ln (5 - x) + \frac{1}{10} \ln (5 + x)$$

$$= \frac{1}{10} \ln \frac{5 + x}{5 - x}$$

$$\frac{1}{25 - x^2} = (25 - x^2)^{-1} = 25^{-1} \left(1 - \frac{x^2}{25}\right)^{-1} = \frac{1}{25} \left[1 - \left(\frac{x}{5}\right)^2\right]^{-1}$$

$$= \frac{1}{25} \left[1 + (-1)\left(-\frac{x^2}{25}\right) + (-1)(-2)\left(-\frac{x^2}{25}\right)^2 \frac{1}{2!} + (-1)(-2)(-3)\left(-\frac{x^2}{25}\right)^3 \frac{1}{3!}\right]$$

$$= \frac{1}{25} + \frac{x^2}{625} + \frac{x^4}{25^3} + \frac{x^6}{25^4} + \dots$$

$$\frac{1}{25 - x^2} = \frac{1}{25} \left(1 + \frac{x^2}{25} + \frac{x^4}{25^2} + \frac{x^6}{25^3} + \dots\right)$$

integrating both sides

$$\int \frac{1}{25 - x^2} \, dx = \frac{1}{10} \ln \frac{5 + x}{5 - x} = \frac{1}{25^2} x + \frac{x^3}{25^2 \times 3} + \frac{x^5}{25^3 \times 5}$$

$$\ln \frac{5 + x}{5 - x} = \frac{10}{25} x + \frac{10}{25^2 \times 3} x^3 + \frac{10}{25^3 \times 5} x^5 + \dots$$

$$= \frac{2}{5} x + \frac{2}{375} x^3 + \frac{2}{15,625} x^5 + \dots$$

Alternatively $\ln \dfrac{5 + x}{5 - x} = \ln (5 + x) - \ln (5 - x)$

$$= \ln 5 (1 + x/5) - \ln 5 (1 - x/5)$$

$$= \ln 5 + \ln (1 + x/5) - \ln 5 - \ln (1 - x/5)$$

214

$$= \ln (1 + x/5) - \ln (1 - x/5), \text{ using}$$

$$\ln (1 + x) = x - \frac{x^2}{2} + \frac{x^3}{3} - \ldots$$

$$\ln (1 + x/5) = x/5 - \frac{(x/5)^2}{2} + \frac{(x/5)^3}{3} - \ldots$$

$$\ln (1 - x/5) = - x/5 - \frac{(x/5)^2}{2} - \frac{(x/5)}{3} - \ldots$$

$$\ln (1 + x/5) - \ln (1 - x/5) = \frac{2x}{5} + \frac{2}{3} (x/5)^3 + \frac{2}{5} (x/5)^5$$

$$\ln \frac{5 + x}{5 - x} = \frac{2}{5}x + \frac{2}{375} x^3 + \frac{2}{15,625} x^5 + \ldots$$

6.
$$\frac{1}{(x^2 + 4)^{1/2}} = (4 + x^2)^{1/2} = 4^{-1/2} \left[1 + (x/4)^2\right]^{-1/2} = \frac{1}{2} \left[1 + \left(\frac{x}{4}\right)^2\right]^{-1/2}$$

$$= \frac{1}{2} \left[1 + \left(- \frac{1}{2}\right)\left(\frac{x}{4}\right)^2 + \left(- \frac{1}{2}\right)\left(- \frac{3}{2}\right)\left(\frac{x}{4}\right)^4 \frac{1}{2} + \ldots\right]$$

provided $- 1 < x/4 < 1$ $\quad \frac{1}{(x^2 + 4)^{1/2}} = \frac{1}{2} - \frac{1}{64} x^2 + \frac{1}{4096} x^4 - \ldots$
integrating both sides

$$\int \frac{1}{(x^2 + 4)^{1/2}} \, dx = \frac{1}{2} x - \frac{1}{64} \times \frac{x^3}{3} + \frac{1}{4096} \times \frac{x^5}{5} - \ldots$$

$$\sinh^{-1} \frac{x}{2} = \frac{1}{2} x - \frac{1}{192} x^3 + \frac{1}{20,480} x^5 -$$

7.
$$\frac{1}{\sqrt{1 - x^2}} = \frac{1}{(1 - x^2)^{1/2}} = (1 - x^2)^{-1/2}$$

$$= \left[1 + \left(- \frac{1}{2}\right)(- x^2) + \left(- \frac{1}{2}\right)\left(- \frac{3}{2}\right)(- x^2)^2 \frac{1}{2} + \ldots\right]$$

$$= 1 + \frac{1}{2} x^2 + \frac{3}{8} x^4 + \ldots$$

integrating both sides

$$\int \frac{1}{\sqrt{1-x^2}} \, dx = x + \frac{1}{6} x^3 + \frac{3}{40} x^5 + \ldots$$

$$\sin^{-1} x = x + \frac{1}{6} x^3 + \frac{3}{40} x^5 + \ldots$$

If $x = 0, c = \cos^{-1} 0 = \pi/2$

$$\cos^{-1} x = \pi/2 - x - \quad \frac{\pi}{2} - x - \frac{1}{6} x^3 - \frac{3}{40} x^5 -$$

Alternatively

$f(x) = \cos^{-1} x$

8.   (i) $f(x) = \sinh^2 x, \ f'(x) = 2 \sinh x \cosh x$
$f''(x) \ 2 \cosh^2 x + 2 \sinh^2 x$
$f'''(x) = 4 \sinh x \cosh x + 4 \sinh x \cosh x$
$f'''(x) = 8 \sinh x \cosh x$
$f^{iv}(x) = 8 \cosh^2 x + 8 \sinh^2 x$
$f^{v}(x) = 16 \cosh x \sinh x + 16 \sinh x \cosh x$
    $= 32 \sinh x \cosh x$
$f(0) = 0 \quad f'(0) = 0, f''(0) = 2, f'''(0) = 0$
$f^{iv}(0) = 8, f^{v}(0) = 0 \quad f^{vi}(x) = 32 \cosh^2 x \ 32 \sinh^2 x$
$f^{vi}(0) = 32$

$$\sinh^2 x = x^2 + \frac{1}{3} x^4 + \frac{2}{45} x^6 + \ldots$$

(ii) $f(x) = \log_e (1 - x^2), f'(x) = \dfrac{1}{1-x^2} \times (-2x) = -2x(1-x^2)^{-1}$

$f''(x) = -2(1-x^2)^{-1} - 2x(1-x^2)^{-2}(-1)(-2x)$
    $= -2(1-x^2)^{-1} - 4x^2(1-x^2)^{-2}$
$f'''(x) = 2(1-x^2)^{-2}(-2x) - 8x(1-x^2)^{-2} - 8x^2(1-x^2)^{-3}(-2x)$
    $= 4x(1-x^2)^{-2} - 8x(1-x^2)^{-2} - 16x^3(1-x^2)^{-3}$
$f^{iv}(x) = 8x(1-x^2)^{-3}(-2x) - 4(1-x^2)^{-2} - 8(1-x^2)^{-2}$
    $+ 16x(1-x^2)^{-3}(-2x)$
      $- 48x^2(1-x^2)^{-3} + 48x^3(1-x^2)^{-4}(-2x)$
    $= -16x^2(1-x^2)^{-3} - 12(1-x^2)^{-2} - 32x^2(1-x^2)^{-3}$
  $- 48x^2(1-x^2)^{-3} - 96x^4(1-x^2)^{-4}$
      $= -12(1-x^2)^{-2} - 96x^2(1-x^2)^{-3} - 96x^4(1-x^2)^{-4}$
$f^{v}(x) = 24(1-x^2)^{-3}(-2x) - 192x(1-x^2)^{-3} + 288x^2(1-x^2)^{-4}(-2x)$

$$+ 384\, x^4 (1 - x^2)^{-4} (- 2x) - 384\, x^3 (1 - x^2)^{-4}$$
$$= - 48x (1 - x^2)^{-3} - 192x (1 - x^2)^{-3} - 576\, x^3 (1 - x^2)^{-4}$$
$$- 768\, x^5 (1 - x^2)^{-4} - 384\, x^3 (1 - x^2)^{-4}$$
$$= - 240\, x (1 - x^2)^{-3} - 960\, x^3 (1 - x^2)^{-4} - 768\, x^5 (1 - x^2)^{-4}$$
$$f^{\,vi} (x) = - 240 (1 - x^2)^{-3} - 240\, x (- 3) (1 - x^2)^{-4} (- 2x)$$
$$- 960 \times 3\, x^2 (1 - x^2)^{-4} - 960\, x^3 (- 4) (1 - x^2)^{-5} (- 2x)$$
$$- 768 \times 5\, x^4 (1 - x^2)^{-4} - 768\, x^5 (- 4) (1 - x^2)^{-5} (- 2x)$$
$$f (0) = 0, f^{\,\prime} (0) = 0, f^{\,\prime\prime} (0) = - 2, f^{\,\prime\prime\prime} (0) = 0, f^{\,iv} (0) = - 12$$
$$f^{\,v} (0) = 0 \quad f^{\,vi} (0) = - 240$$

$$\log_e (1 - x^2) = - x^2 - \frac{x^4}{2} - \frac{x^6}{3}.$$

Alternatively, applying the expansion

$$\log_e (1 - x) = - x - \frac{x^2}{2} - \frac{x^3}{3} - \frac{x^4}{4} - \ldots$$

$$\log_e (1 - x^2) = - x^2 - \frac{x^4}{2} - \frac{x^6}{3} - \frac{x^8}{4} - \ldots$$

a much easier approach.

(iii) $\quad f (x) = \sin^2 x, f^{\,\prime} (x) = 2 \sin x \cos x$
$$f^{\,\prime\prime} (x) = 2 \cos^2 x + 2 \sin x (- \sin x) = 2 \cos^2 x - 2 \sin^2 x$$
$$f^{\,\prime\prime\prime} (x) = 4 \cos x (- \sin x) - 4 \sin x \cos x = - 8 \sin x \cos x$$
$$f^{\,iv} (x) = - 8 \cos^2 x + 8 \sin^2 x$$
$$f^{\,v} (x) = - 16 \cos x (- \sin x) + 16 \sin x \cos x$$
$$= 32 \sin x \cos x$$
$$f^{\,vi} (x) = 32 \cos^2 x - 32 \sin^2 x$$
$$f (0) = 0, f^{\,\prime} (0) = 0, f^{\,\prime\prime} (0) = 2, f^{\,\prime\prime\prime} (0) = 0, f^{\,iv} (0) = - 8$$
$$f^{\,v} (0) = 0, \quad f^{\,vi} (0) = 32$$

$$\sin^2 x = x^2 - \frac{1}{3} x^4 + \frac{2}{45} x^6$$

(iv) $\quad f (x) = \cos^2 x, f^{\,\prime} (x) = - 2 \sin x \cos x, f^{\,\prime\prime} (x) = - 2 \cos^2 x + 2 \sin^2 x$
$$f^{\,\prime\prime\prime} (x) = - 4 \cos x (- \sin x) + 4 \sin x \cos x = 8 \sin x \cos x$$
$$f^{\,iv} (x) = 8 \cos^2 x - 8 \sin^2 x, f^{\,v} (x) = 16 \cos x (- \sin x) - 16 \sin x \cos x$$
$$f^{\,v} (x) = - 32 \sin x \cos x, \quad f^{\,vi} (x) = - 32 \cos^2 x + 32 \sin^2 x$$
$$f (0) = 1, f^{\,\prime} (0) = 0, f^{\,\prime\prime} (0) = - 2, f^{\,\prime\prime\prime} (0) = 0, f^{\,iv} (0) = 8$$

$f^v(0) = 0, f^{vi}(0) = -32.$

$$\cos^2 x = 1 - x^2 + \frac{1}{3}x^4 - \frac{2}{45}x^6.$$

Observe that the latter could have been derived from $\sin^2 x = x^2$

$$\sin^2 x = x^2 - \frac{1}{3}x^4 + \frac{2}{45}x^6 \text{ if } \sin^2 x = 1 - \cos^2 x = x^2 - \frac{1}{3}x^4 + \frac{2}{45}x^6$$

therefore, $\cos^2 x = 1 - x^2 + \frac{1}{3}x^4 - \frac{2}{45}x^6.$

9. Taylor's theorem states:-

$$f(x+h) = f(x) + h f'(x) + \frac{h^2}{2!}f''(x) + \frac{h^3}{3!}f'''(x) + \dots$$

$$\tan 45° \, 2' = \tan\left(45 + \frac{2}{60}\right)^\circ = \tan\left(45 + \frac{1}{30}\right)^\circ$$

$$= \tan\left(\frac{\pi}{4} + \frac{\pi}{30 \times 180}\right) = \tan\left(\frac{\pi}{4} + \frac{\pi}{5,400}\right)$$

$$\tan\left(\frac{\pi}{4} + \frac{\pi}{5,400}\right) = \tan\frac{\pi}{4} + \frac{\pi}{5,400}\sec^3\frac{\pi}{4} + \left(\frac{\pi}{5,400}\right)^2\frac{1}{2} \, 2\sec^2\frac{\pi}{4}\tan\frac{\pi}{4}$$

$$+ \left(\frac{\pi}{5,400}\right)^3\frac{1}{3!} \, 4\sec^2\frac{\pi}{4}\tan^2\frac{\pi}{4} + \dots = 1 + \frac{\pi}{5,400}(2) + \frac{\pi^2}{5,400^2} \times \frac{1}{2} \times 2 \, (2) \times 1$$

$$+ \frac{\pi^3}{5,400^3} \times \frac{1}{6} \times 4 \, (2) \, (1) + \dots \qquad = 1 + 1.1635528\text{x}\,10^{-3} + 6.769276\text{x}\,10^{-7}$$

$$+ 2.6254701\text{x}\,10^{-10} \qquad\qquad = 1.0011642$$

From the calculator $\tan 45° \, 2' = 1.0011642$ which agrees with the answer.

(ii) $\log_e 1.001 = \log_e (1 + 0.001)$

$$\log_e (1 + 0.001) = \log_e 1 + 0.001 \times \frac{1}{1 + 0.001} + \frac{0.001^2}{2}\left(-\frac{1}{(1 + 0.001)^2}\right)$$

$$+ \frac{0.001^3}{3!} \times \frac{2}{(1 + 0.001)^3} + \dots$$

$$= 9.99 \times 10^{-4} - 4.990015 \times 10^{-7} + 3.3233533 \times 10^{-10} = 9.985 \times 10^{-4}.$$

From the calculator $\ln 1.001 = 9.9950033 \times 10^{-4}$ which agrees to five significant figures.

10. $\log_e (1 + x + x^2) = \log_e \frac{1 - x^3}{1 - x}$ using $\log_e (1 + x) = x - \frac{x^2}{2} + \frac{x^3}{3} - \frac{x^4}{4} + \dots$

and $\log_e (1 - x) = x - \frac{x^2}{2} - \frac{x^3}{3} - \frac{x^4}{4} - \dots$

$\log_e (1 - x^3)/(1 - x) = \log_e (1 - x^3) - \log_e (1 - x)$

$$= \left(-x^3 - \frac{x^6}{2} - \dots \right) - \left(-\frac{x^2}{2} - \frac{x^3}{3} - \frac{x^4}{4} - \frac{x^5}{5} - \frac{x^6}{6} - \dots \right)$$

$$= x + \frac{x^2}{2} - \frac{2}{3} x^3 + \frac{x^4}{4} + x^5 - \frac{1}{3} x^6 - \dots$$

$\ln 1.0204 = \ln (1 + x + x^2)$ where $x = 0.02$

$$= 0.02 + \frac{0.0004}{2} - \frac{2}{3} 0.000008 + \frac{0.00000016}{4}$$

$$+ \frac{0.0000000032}{5} - \frac{1}{3} 0.000000000064$$

$$= 0.020195$$

From the calculator $\ln 1.0204$ is correct to seven significant figures.

11.   (i) $\ln (1 - 4x) = -4x - \frac{(4x)^2}{2} - \frac{(4x)^3}{3} - \frac{(4x)^4}{4} - \dots$

$$= -4x - 8x^2 - \frac{64}{3} x^3 - 16 x^4 - \dots$$

    (ii) $\ln (3 + 5x) = \ln 3 \left(1 + \frac{5}{3} x \right) = \ln 3 + \ln \left(1 + \frac{5}{3} x \right)$

$$= \ln 3 + \frac{4}{3}x - \left(\frac{5}{3}x\right)^2 \frac{1}{2} + \left(\frac{5}{3}\right)^3 \frac{1}{3} - \ldots$$

(iii)
$$\ln (2 + 7x) = \ln 2 \left(1 + \frac{7x}{2}\right) = \ln 2 + \ln \left(1 + \frac{7x}{2}\right)$$

$$= \ln 2 + \frac{7x}{2} - \left(\frac{7x}{2}\right)^2 \frac{1}{2} + \left(\frac{7x}{2}\right)^3 \frac{1}{3} - \left(\frac{7x}{2}\right)^4 + \ldots$$

The ranges of values of x for which the expansions are valid are:-

(i) $\quad -\dfrac{1}{4} \le x < \dfrac{1}{4}$ (ii) $\quad -\dfrac{3}{5} < x \le \dfrac{3}{5}$ (iii) $\quad -\dfrac{2}{7} < x \le \dfrac{2}{7}.$

12. From (3) $\quad \tan^{-1} x = x - \dfrac{1}{3} x^3 + \dfrac{1}{5} x^5 - \ldots$

If $x = 1$, $\tan^{-1} = \pi/4$

$$\frac{\pi}{4} = 1 - \frac{1}{3}(1)^3 + \frac{1}{5}(1)^5 - \ldots \qquad \frac{\pi}{4} = 1 - \frac{1}{3} + \frac{1}{5} - \frac{1}{7} + \ldots$$

$$\pi = 4\left(1 - \frac{1}{3} + \frac{1}{5} - \frac{1}{7} + \ldots\right).$$

13. From (2) $\sin^{-1} x = x + \dfrac{1}{6} x^3 + \dfrac{3}{40} x^5 + \ldots$

If $\quad x = \dfrac{1}{2}$, $\sin^{-1} \dfrac{1}{2} = \pi/6$

$$\pi/6 = \frac{1}{2} + \frac{1}{6}\left(\frac{1}{2}\right)^3 + \frac{3}{40}\left(\frac{1}{2}\right)^5 + \ldots$$

$$\pi = 6\left(\frac{1}{2} + \frac{1}{2 \times 3 \times 2^3} + \frac{3 \times 1}{2 \times 4 \times 5 \times 2^5} + \ldots\right)$$

14. $\quad\int \dfrac{1}{\sqrt{1-x^2}}\, dx \quad$ let $x = \cos\Theta \quad \dfrac{dx}{d\Theta} = -\sin\Theta$

$$= \int \frac{1 \times (1-\sin\Theta\,d\Theta)}{\sqrt{1-\cos^2\Theta}} = -\int d\Theta = -\cos^{-1}x + c$$

$$(1-x^2)^{-1/2} = 1 + \left(-\frac{1}{2}\right)(-x^2) + \left(-\frac{1}{2}\right)\left(-\frac{3}{2}\right)(-x^2)^2\,\frac{1}{2}$$

$$+ \left(-\frac{1}{2}\right)\left(-\frac{3}{2}\right)\left(-\frac{5}{2}\right)\left(-x^2\right)^3\,\frac{1}{6}$$

$$= 1 + \frac{1}{2}x^2 + \frac{3}{8}x^4 + \frac{5}{16}x^6 + \dots$$

integrating both sides $\quad c - \cos^{-1}x = x + \dfrac{1}{6}x^3 + \dfrac{3}{40}x^5 + \dots$

MISCELLANEOUS

1. A curve is defined parametrically by $x = \dfrac{2t}{1+t}$, $y = \dfrac{t^2}{1+t}$.

Prove that the normal to the curve at the point $(1, 1/2)$ has equation $6y + 4x = 7$ (5 marks)

Determine the coordinates of the other point of intersection of this normal with the curve (4 marks)

AEB Nov. 1990

2. In Fig. 49, CPB is a straight road running alongside a flat field where **CB** is 60 metres and **P** lies between C and B. The point A is in the field and C is the point on the road closest to A so that AC is 15 metres. A trench is to be dug from A to B in two straight section AP and PB.

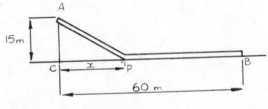

Fig. 49

The cost of digging in the field is £50 per metre and along the road £40 per metre. When $PC = x$ metres the total cos of digging is £y. Shows that

$$y = 50\sqrt{225 + x^2} + 40(60 - x).$$ (3 marks)

(a) Find the positive value of x for which y has a stationary value and determine whether it is a maximum or miniumum value. (7 marks)

(b) Sketch the graph of y against x for $0 \le x \le 60$ and state the greatest and least costs of digging the trench. (6 marks)

AEB Nov. 1990

3. (a) A curve has equation $y = (5 - 3x)^{-2}$. Find the equation of the the tangent to the curve at the point $\left(1, \dfrac{1}{4}\right)$.

(b) Given that $xy = 3x^2 + y^2$, find $\dfrac{dy}{dx}$ giving your answer in terms of x and y.

AEB 1990 June.

4. Given that $e^y = e^x + e^{-x}$, show that

$$\frac{d^2y}{dx^2} + \left(\frac{dy}{dx}\right)^2 - 1 = 0.$$

Find the values of y, $\dfrac{dy}{dx}$ and $\dfrac{d^2y}{dx^2}$ when $x = 0$.

By further differentiation, evaluate $\dfrac{d^3y}{dx^3}$ and $\dfrac{d^4y}{dx^4}$ when $x = 0$.

Use Maclaurin's series to show that $y = \ln 2 + \dfrac{1}{2} x^2 - \dfrac{1}{12} x^4$

when x is small enough for x^5 and higher powers of x to be neglected.

The finite region R is bounded by the curve $e^y = e^x + e^{-x}$, the

x-axis and the lines $x = \pm \dfrac{1}{2}$. The region is rotated completely

about the x-axis to form a solid of revolution. Using the first two terms in the series expansion of y in ascending powers of x, find an estimate of the volume of this solid, giving your answer to <u>one</u> decimal place. (4 marks)

AEB Nov. 1989.

5. Given that $y = \sin x + \dfrac{1}{2} \sin 2x + \dfrac{1}{3} \sin 3x$, show that

$\dfrac{dy}{dx} = (1 + 2 \cos x)^2 \cos 2x$. (4 marks)

Find the complete set of values of x for which $\dfrac{dy}{dx} < 0$ in the interval $0 < x < \pi$. (4 marks)

AEB Nov. 1989

6.   The highest point **H** of a small island is h metres above sea level. A vertical television mast **TH** of height a metres is placedd with its base at **H**. The point **O** is at sea level directly beneath the mast. A small boat **P** is anchored at a distance x metres measured horizontally from **O** and angle **TPH** = ⊙ (see Fig. 50).

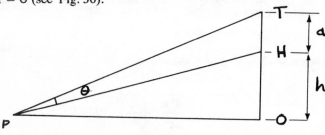

Fig. 50

Show that $\tan \Theta = \dfrac{ax}{x^2 + ah + h^2}$.   (4 marks)

As x varies, find, in terms of a and h, the positive value of x for which

$$\frac{d}{dx}\left(\frac{ax}{x^2 + ah + h^2}\right) = 0$$ (3 marks)

Prove that this value of x gives a maximum value of ⊙.   (3 marks)

The point **P** now moves in a straight line directly away from **0** with constant speed
12 ms^{-1}. Given that $h = 20$ and $a = 10$, find, in radians per second to two significant figures, the rate of change of ⊙ at the instant
when $x = 50$ (6 marks)

AEB Nov. 1989.

7.   The radius r cm of a circular ink spot, t seconds after it first appears, is

given by $r = \dfrac{1 + 4t}{2 + 1}$

Calculate

(a)   The time taken for the radius to double its initial size   (2 marks)

(b)   the rate of increase of the radius in cm s^{-1}
         when $t = 3$ (4 marks)

224

(c)   the value to which r tends as t tends to infinity.   (2 marks)

AEB June 1989

8.   Differentiate with respect to t

     (i) $e^{2t} \cos t$ [2]

     (ii) $\sin (t^3 + 4)$.   [3]

     (iii) $\dfrac{t}{t^2 + 1}$ C.N.89 [3]

9.   It is given $\dfrac{dy}{dx} = y + 2x$ and that $y = 1$ when $x = 0$. Show that $\dfrac{d^2 y}{dx^2} = 3$ when $x = 0$.

Find the first three terms in the Maclaurin series for y.

C.N.89 [5]

10.   Find the root of the equation

$$e^{2 - 2x} = 2e^{-x},$$

giving your answer exactly, in terms of logarithms.   [3]

Show that the curve

$$y = e^{2 - 2x} - 2e^{-x}$$

has a turning point at $(2, -e^{-2})$.   [4]

Sketch the curve for $x \geq 0$.   [3]

Hence state the set of values of k for which the equation

$$e^{2 - 2x} - 2e^{-x} = k$$

has two distinct positive roots.   C.N.89 [2]

11.   Find the equation of the normal to the curve

225

$$3y^2 - x^2 = 3$$

at the point (3.2).   C.N.89[5]

12.

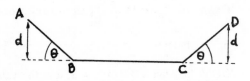

Fig. 51

A length of channel of given depth d is to be made from a rectangular sheet of metal of width $2a$. The metal is to be bent in such a way that the cross-section $ABCD$ is as shown in the figure, with $AB + BC + CD = 2a$ and with AB and CD each inclined to the line BC at an angle Θ.

Show that $BC = 2\,(a$ - cosec $\Theta)$ and that the area of the cross-section $ABCD$ is

$$2ad + d^2 \text{ (cot } \Theta \text{ - 2 cosec } \Theta).$$ [5]

Show that the maximum of $2ad + d^2$ (cot Θ - 2 cosec Θ), as Θ varies, is $d\ (2a - d\sqrt{3})$.

By considering the length of BC, show tht the cross-sectional area can only be made equal to this maximum value if $2d \le a\ \sqrt{3}$

                                                                       C.N.89[2]

13.   By considering the graphs of $y = 1 + x$ and $y = ke^x$, or otherwise, show the equation $e^{-x}\,(1 + x) = k$ has exactly two real roots for x when k is a constant such that $0 < k < 1$. State the number of real roots for each of the cases $k < 0$ and $k > 1$.   [5]

For the case $k = 0.1$, the negative root of the equation is α and the positive root is β. Determine the integer which is closest to α and the integer which is closest to β. [3]

For the case $k = 0.2$, use any suitable method to find, correct to four significant figures, the root of the equation which is close to 3. C.N.89 [6]

14. (i) Given that $y = \ln \dfrac{x}{x^2}$, find the value of $\dfrac{dy}{dx}$ when $x = e$.

(ii) Given that $\tan x + \tan y = 3$, find the value of $\dfrac{dy}{dx}$ when $x = \pi/4$. C.J.90 [7]

15. Find the coordinates of the turning points on the curve with equation $y = x (x - 1)^2$ and sketch the curve. [5]

(i) Sketch the curve $y^2 = x (x - 1)^2$. [2]

Find the set of real values of k such that the equation $x (x - 1)^2 = k^2$ has exactly one real root. [2]

(ii) For the equation $x (x - 1)^2 = 1\,000\,000$, find the integer n such that the interval $[n - 1, n]$ contains the root. Taking n as first approximation, use the Newton-Raphson formula once to obtain a second approximation. C. [5]

16. Let $f (x) = e^{-x} \sin x$. Show that

$$f'' (x) = - 2 \{f' (x) + f(x)\}.$$ [4]

By further differentiation of this result, or otherwise, find the value of $f^{(5)} (0)$, and write down the series expansion of $f (x)$ in ascending powers of x up to and including the term in x^5. [5]

17. [In this question, take the value of g to be 10 ms^{-2}.]

A bead of mass 0.4 is threaded on a horizontal straight wire. The

coefficient of friction between the bead and the wire is $\dfrac{3}{4}$. At time t seconds, where $0 \le t \le 4$.

227

$$F = t(4 - t).$$

Find the time at which the bead begins to move. [4]

At time t seconds the velocity of the bead is v m s^{-1}. Show that, provided $1 \le t \le 4$.

(i) $\qquad \dfrac{dv}{dt} = \dfrac{5}{2}(t - 1)(3 - t),$ [4]

(ii) $\qquad v = \dfrac{5}{6}(t - 1)^2(4 - t).$ [4]

June 1989 C

Find the rate of working of the force F when $t = 3$.

18.　　(a)　　Show that the tangent at the point $(e, 1)$ to the graph $y = \ln x$ passes through the origin, and deduce that the line $y = mx$ cuts the graph $y = \ln x$ in two points provided that $\;0 < m < \dfrac{1}{e}.$ [6]

　　　　(b)　　For each root of the equation $\ln x = \dfrac{1}{3}x$ find an integer n such that the interval $n < x < n + 1$ contains the root. Using linear interpolation, based on $x = n$ and $x = n + 1$, find a first approximation to the smaller root, giving one place of decimals in your answer. [5]

　　　　　　Using your first approximation, obtain, by the Newton-Raphson method, a second approximation to the smaller root, giving two places of decimals in your answer. [3]

June 1989C

19.　　For each of the following curves, find the gradient at the specified point:-

　　　　(i)　　$x^3 + y^3 = 9$, at the point $(1, 2)$;
　　　　(ii)　　$y = 2^x$, at the point (0.1). [6]

June 1989(8) C

20.　　The equation of a curve is $y = ax^2 - 2bx + c$, where a, b and c are constants, with $a > 0$.

　　　　(i)　　Find, in terms of a, b and c, the coordinates of the turning point on the curve, and state whether it is a maximum point or a minimum point. [4]

(ii) Given that the turning point of the curve lies on the line $y = x$, find an expression for c in terms of a and b. Show that, in this case, whatever the value of b,

$$c \geq - \frac{1}{4a}.$$ [4]

(iii) Find the numerical values of a, b and c when the curve passes through the point $(0, 6)$ and has a turning point at $(2, 2)$. [4]

June 1989 (12) C

21. (a) The parametric equations of a curve are $x = a \cos^3 \Theta$, $y = a \sin^3 \Theta$, where a is a positive constant and

$$0 < \Theta < \frac{1}{2}\pi.$$

(i) Show that $\dfrac{dy}{dx} = - \tan \Theta$. [2]

(ii) The tangent to the curve at the point with parameter Θ cuts the axes at S and T. Write down the equation of this tangent and show that the distance ST is independent of Θ. [6]

(b) The parametric equations of a curve are $x = at$, $y = at^2$, where a is a constant. The points P (ap, ap^2) and Q (aq, aq^2) lie on the curve. Find and simplify an expression, in terms of p and q, for the gradient of the chord PQ.

Deduce from your expression that the gradient of the tangent to the curve at P is $2p$. [4]

June 1989 (16) C

22. Given that $y = \tan \left(\dfrac{1}{4}\pi + x \right)$, show that $\dfrac{d^2y}{dx^2} = 2y \dfrac{dy}{dx}$ [3]

By repeated differentiation of this result, or otherwisee, find the series expansion of y in ascending powers of x up to and including the term in x^4. [7]

When $x = 0.1$, the value of y, correct to 8 decimal places, is

229

1.223 048 88. Verify that, in this case, the series expansion up to and including the term in x^4 gives and estimate that is in error by about 0.004%. [2]

June 1989 (18) C

23.

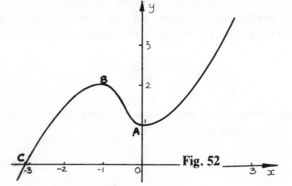

Fig. 52

The diagram shows the graph of $y = f(x)$. The points A, B, C have coordinates $(0, 1)$, $(-1, 2)$, $(-3, 0)$ respectively. Sketch separately the graphs of

(i) $y = f(-x)$,
(ii) $y = f(x + 3)$.

showing in each case the coordinates of the points corresponding to A, B and C.

24.   Differentiate with respect to t

(i) $e^{2t} \cos t$. [2]
(ii) $\sin(t^3 + 4)$. [3]

(iii) $\dfrac{t}{t^2 + 1}$. C. [3]

25.   Find the equation of the normal to the curve

$$3y^2 - x^2 = 3$$

at the point $(3, 2)$. C. [5]

26.   (i)   Given that $y = \dfrac{\ln x}{x^2}$, find the value of $\dfrac{dy}{dx}$ when $x = e$.

(ii)   Given that $\tan x + \tan y = 3$, find the value of

$$\frac{dy}{dx} \text{ when } x = \frac{1}{4}\pi.$$

C. [7]

27.   Find the root of the equation

$$e^{2 - 2x} = 2e^{-x},$$

giving your answer exactly, in terms of logarithms.   [3]

Show that the curve

$$y = e^{2 - 2x} - 2e^{-x}$$

has a turning point at $(2, -e^{-2})$.   [4]

Sketch the curve for $x \geq 0$.   [3]

Hence state the set of values of k for which the equation

$$e^{2 - 2x} - 2e^{-x} = k$$

has two distinct positive roots.   C. [2]

28.   A water tank has the shape of an open rectangular box of length 1 m, width
       0.5 m and height 0.5 m. Water may be drained from the tank through a tap
       at the bottom of the tank, and it is known that, when the tap is open, water
       leaves at a rate of 100 h litres per minute, where h m is the depth of water
       in the tank. When the tap is open, water is also fed into the tank at a
       constant rate of 50 litres per minute and no water is fed into the tank when
       the tap is closed. Show that, t minutes after the tap has been opened, the
       variable h satisfies the differential equation

$$10 \, \frac{dh}{dt} = 1 - 2h.$$

On a particular occasion, the tap was opened when $h = 0.25$ and closed
when $h = 0.375$. Show that the tap was opened for $5 \ln 2$ minutes.

June 1984 P2 (13) U. L

29.   Show that $\dfrac{d}{dr} \left[\sin^{-1} (a/r) \right] = - \dfrac{a}{r \sqrt{(r^2 - a^2)}}.$

A rectangular field has sides of length 10 m and 20 m. A goat is tethered
to a corner of the field by an inelastic rope of length r m where $10 < r <
20$. Show that the goat has access to an area A m^2 of the field, where

$$A = 5 \sqrt{(r^2 - 100)} + \frac{1}{2} r^2 \sin^{-1} (10/r).$$

Show that $\dfrac{dA}{dr} = r \sin^{-1} (10/r).$

Apply the Newton-Raphson procedure once to the equation $A - 100 = 0$, with starting value of $r = 0$, to show that the goat has access to one half of the area of the field when r is approximately equal to 11.4.

June 1984 P2 (14) U. L.

30.   A right circular cylinder is to be cut from a right circular solid cone, of height H and base radious R. The axis of the cylinder lies along the axis of the cone. The circumference of one end of the cylinder is in contact with the curved surface of the cone and the other end of the cylinder lies on the base of the cone. Show that V, the volume of the cylinder, is given by

$$V = \frac{\pi H x^2 (R - x)}{R},$$

where x is the radius of the cylinder. Show also that, as x varies, the maximum possible value of V is $4\pi R^2 H/27$.   January 1985 (14) U. L.

Jan. 1985 P3 (2)

31.   A curve is given parametrically by $x = \sin t, y = \cos^3 t, -\pi < 1 \le \pi$. Show that (a) $-1 \le x \le 1$ and $-1 \le y \le 1$,

(b) $\dfrac{dy}{dx} = a \sin 2t$, where a is constant, and given the value of a.

Find the value of $\dfrac{dy}{dx}$ when $x = 0$ and show that the curve has points of

inflexion where $t = -3\pi/4, -\pi/4, \pi/4$ and $3\pi/4$. Sketch the curve.

January 1985 P3 (2) U. L.

32.   Find an equation of the tangent to the curve $y = e^x$ at the point with coordinates (I, e) and deduce that this tangent passes through the origin. Hence show graphically that, when $m > e$, the equation $mx = e^x$ has exactly two distinct roots.

Given that $f(x) = 4x - e^x$, show that the equation $f(x) = 0$ has a root in the interval $0.3 \le x \le 0.4$.

Taking $x = 0.35$ as an initial approximation and using the Newton-Raphson procedure, obtain two further approximations to the root of the equation $f(x) = 0$, giving your answers to 3 decimal places.

January 1986 P2 (14) U. L.

33.   A right pyramid has square horizontal base of side x and the vertex of the pyramid is at height y vertically above the centre of the base. Given that the total surface area of the pyramid is a^2, where a is a constant find an equation relating x, y and a.

Hence show that the volume V of the pyramid is given by

$$36\ V^2 = a^2\ x^2\ (a^2 - 2\ x^2).$$

Given that x and y vary, find in terms of a, the greatest possible volume of the pyramid. Justify heat the volume you have found is a maximum.

June 1986 Special (9) U. L.

34.   Show that the volume V of a right circular cone of slant height l and semi-vertical angle Θ, where $0 < \Theta < \pi/2$, is given by

$$V = \frac{1}{3} H\ l^3\ \sin^2 \Theta \text{ as } \Theta$$

Given that l is constant and that Θ varies, find

(a)   the value of 0 for which V is a maximum,

(b)   the maximum value of V in terms of l and π.

Jan 1987 P2   (11)   **U. L.**

35.   A vertical wall, 2.7 m high, runs parallel to the wall of a house and is at a horizontal distance of 6.4 m from the house. An extending ladder is placed to rest on the top **B** of the wall with one end **C**, against the house and the other end **A** resting on horizontal ground, as shown in Fig. 53.

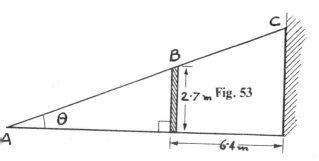

Fig. 53

The points, **A, B** and **C** are in a vertical plane at right angles to the wall and the ladder makes an angle Θ, where $0 < Θ < π/2$, with the horizontal.

Show that the length y metres, of the ladder is given by $y = \dfrac{2.7}{\sin Θ} + \dfrac{6.4}{\cos Θ}$.

As Θ varies, find the value of tan Θ for which y is a minimum. Hence find the minimum value of y.

June 1987 P2 (11) U.L.

36.

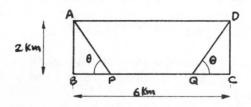

A

D

2 km

θ

θ

B   P

Q   C

6 km

Fig. 54

The diagram shows a house at A, a school at D and a straight canal $B\ C$, where $ABCD$ is a rectangle with $AB = 2$ km and $BC = 6$ km.

During the winter, when the canal freezes over, a boy travels from A to D by walking to a point P on the canal, skating along the canal to a point Q and then walking from Q to D. The points P and Q being chosen so that the angles APB and DQC are both equal to Θ.

Given that the boy walks at a constant speed of 4 km h^{-1} and skates at a constant speed of 8 km h^{-1}, show that the time, T minuts, taken for the boy to go from A to D along this route is given by

$$T = 15 \left(3 + \frac{4}{\sin Θ} - \frac{2}{\tan Θ} \right).$$

Show that, as Θ varies, the minimum time for the journey is approximately 97 minutes.   U. L. Janury 1988 P2 (4) (11 marks)

37.   In Fig. 55, the chord AB, of a circle of radius 2 cm, divides the sector $CADB$ into two portions such that the area of triangle ABC is three quarters of the area of the sector. Given that angle ACB is 2Θ, show that 2 sin 2Θ = 3Θ.

234

Show that a root of this equation lies between 0.63 and 0.65. Using 0.63 as a first approximation to this root, together with a single application of the Newton-Raphson procedure, find a second approximation to this root giving your answer to 3 significant figures.

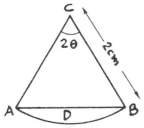

U. L. June 1988 (12)

Fig. 55

38.   Sketch the curve $y = (2x - 1)^2 (x + 1)$, showing the coordinates of

(a)   the points where it meets the axes

(b)   the turning points

(c)   the point of inflexion.

By using your sketch, or otherwise, sketch the graph of $y = \dfrac{1}{(2x - 1)^2 (x + 1)}$.

Show clearly the coordinates of any turning points.

U. L. June 1988 (14)

39.

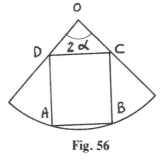

Fig. 56

Fig. 56 shows a rectangle $ABCD$ inscribed symmetrically in a sector of a circle of fixed radius a and fixed angle 2α. Given that $OD = x$ and $AD = BC = y$, show that $a^2 = x^2 + 2xy \cos \alpha + y^2$.

Determine dy/dx in terms of x, y and α and show that the maximum area of the rectangle $ABCD$, as x and y vary, is $a^2 \tan (\alpha/2)$.

U. L. June 1988 Special paper A (4)

40. (i) Differentiate with respect to x

      (a) $e^{3x} \sin(\pi x)$

      (b) $\ln\left[\dfrac{x^2 + 1}{\sqrt{x}}\right]$.

    (ii) A coloured liquid is poured onto a large flat cloth and forms a circular sain, the area of which grows at a steady rate of $3 \text{ cm}^3 \text{ s}^{-1}$. Calculate, in terms of π,

      (a)   the radius in cm, of the stain 6 s after the stain commences.

      (b)   the rate, in cm s^{-1}, of increase of the radius of the stain at this instant.

U. L. January 1989 (P2) (13)

41. (a) Sketch, for $0 < x < \dfrac{\pi}{2}$, the curve $y = \tan x$. By using your sketch show that the equation $\tan x = \dfrac{1}{x}$ has one and only one root in $0 < x < \dfrac{\pi}{2}$.

    (b) Show further that this root lies between 0.85 and 0.87.

    (c) Taking 0.85 as a first approximation to this root, use the Newton-Raphson procedure once to determine a second approxmation, giving your answer to 4 decimal places.

(d) Show that the curve with equation $y = x \cos x$ has a maximum value when

    $\tan x = \dfrac{1}{x}$ and $0 < x < \dfrac{\pi}{2}$.

(e) Find the area of the finite region bounded by the curve with equation $y = $

    $x \cos, \ \ 0 \le x \le \dfrac{\pi}{2}$, and the x-axis.

U. L. June 1989 (P1) (15)

. Given that $f(\Theta) = \Theta - \sqrt{(\sin \Theta)}$, $0 < \Theta < \frac{1}{2}\pi$, show that

(a) the equation $f(\Theta) = 0$ has a root lying between $\frac{1}{4}\pi$ and $\frac{3}{10}\pi$,

(b) $f'(\Theta) = \dfrac{\cos \Theta}{2\sqrt{(\sin \Theta)}}$.

(c) Taking $3\pi/10$ as a first approxmation to this root of the equation $F(\Theta) = 0$, use the Newton-Raphson procedure one to find a second approximation, giving your answer to 2 decimal places.

(d) Show that $f'(\Theta) = 0$ when $\sin \Theta = \sqrt{5} - 2$.

U. L. January 1990 P1 (13)

43. A sector S of a circle, of radius R, whose angle at the centre of the circle is ϕ radians, is rolled up to form the curved surface of a right cone standing on a circular base. The semi-vertical angle of this cone is Θ radians. Express ϕ in terms of $\sin \Theta$ and show that the volume V of the cone is given by $3V = \pi R^3 \sin^2 \Theta \cos \Theta$. (5 marks)

If R is constant and Θ varies, find the positive value of $\tan \Theta$ for which $\dfrac{dV}{d\Theta} = 0$.

Show further that when this value of $\tan \Theta$ is taken, the maximum value of V is obtained. (6 marks)

Hence show that the maximum value of V is $\dfrac{2\pi R^3 \sqrt{3}}{27}$ and find, in terms of R, the area of the sector S in this case. (5 marks)

AEB J.88

44. Given that $\sinh y = x$, show that $y = \ln\left[x + \sqrt{(1 + x^2)}\right]$.

(4 marks)

Show also that $(1 + x^2)\left(\dfrac{dy}{dx}\right)^2 = 1$. (2 marks)

Sketch the curve with equation $x = \sinh y$ and state the gradient of this curve at the origin. (2 marks)

By further differentiation of the relation $(1 + x^2)\left(\dfrac{dy}{dx}\right)^2 = 1$,

show that

$$(1 + x^2)\dfrac{d^3y}{dx^3} + 3x\dfrac{d^2y}{dx^2} + \dfrac{dy}{dx} = 0.$$

237

Given that x is small, find, in terms of x, a cubic polynomial approximation for

$$\ln\left[x + \sqrt{(1 + x^2)}\right]$$

(4 marks)

Given that x is small, find, in terms of x, a cubic polynomial approximation

for $\ln\left[x + \sqrt{(1 + x^2)}\right]$.

(3 marks)

AEB J.88 P2 (5)

INDEX

DIFFERENTIAL CALCULUS AND APPLICATIONS

Angle between two lines   83, 84
Approximation rates   128, 129

Circular functions   25
Cycloid   67, 68

Differentiation from first principles of
algebraic functions   3, 4
    of trigonometric functions   25
    of exponential functions   42
Derivative of a difference   9
    of an implicit function   18, 35
    of a product function   11
    of a quotient function   14
    of a sum function   9
    of a function of a
        function   16, 35, 56
    of tan x, cot x 26
    cosec x, sec x 27

Equations of normals and
tangents   89
Exponential functions   40

Function of a function   35, 56

Higher derivatives of a
function   114
Hyperbolic functions   54
    graphs   62

Improvement of an approximation 104
Inverse hyperbolic functions   59
    graphs   62
Inverse trigonometric function   35

Leibnitz's theorem   126
L'Hôpital's rule   93
Limit (the concept)   2, 94
Logarithmic functions   48

Maclaurin's expansion   114

Newton-Raphson's method   104
Normals   83
Notation of a gradient   3
Numerical methods for the solution of
differential equations   128

Parametric equations   65
Polynomial approximations   using
Taylor's series   128
Power series   114

Rate of change   76, 98

Second derivatives   76
Small increments   71

Tangents   83
Taylor series   128
Trigonometric functions   85